Progress in the Chemistry of Organic Natural Products

The volumes of this classic series, now referred to simply as "Zechmeister" after its founder, Laszlo Zechmeister, have appeared under the Springer Imprint ever since the series' inauguration in 1938. It is therefore not really surprising to find out that the list of contributing authors, who were awarded a Nobel Prize, is quite long: Kurt Alder, Derek H.R. Barton, George Wells Beadle, Dorothy Crowfoot-Hodgkin, Otto Diels, Hans von Euler-Chelpin, Paul Karrer, Luis Federico Leloir, Linus Pauling, Vladimir Prelog, with Walter Norman Haworth and Adolf F.J. Butenandt serving as members of the editorial board.

The volumes contain contributions on various topics related to the origin, distribution, chemistry, synthesis, biochemistry, function or use of various classes of naturally occurring substances ranging from small molecules to biopolymers.

Each contribution is written by a recognized authority in the field and provides a comprehensive and up-to-date review of the topic in question. Addressed to biologists, technologists, and chemists alike, the series can be used by the expert as a source of information and literature citations and by the non-expert as a means of orientation in a rapidly developing discipline.

All contributions are listed in PubMed.

More information about this series at https://link.springer.com/bookseries/10169

Progress in the Chemistry of Organic Natural Products

Volume 117

A. Douglas Kinghorn · Heinz Falk ·
Simon Gibbons · Yoshinori Asakawa · Ji-Kai Liu ·
Verena M. Dirsch
Editors

Antimalarial Natural Products

By David G. I. Kingston and Maria Belen Cassera

Springer

Editors
A. Douglas Kinghorn
College of Pharmacy
Ohio State University
Columbus, OH, USA

Heinz Falk
Institute of Organic Chemistry
Johannes Kepler University
Linz, Austria

Simon Gibbons
School of Pharmacy
University of East Anglia
Norwich, UK

Yoshinori Asakawa
Faculty of Pharmaceutical Sciences
Tokushima Bunri University
Tokushima, Japan

Ji-Kai Liu
School of Pharmaceutical Sciences
South-Central University for Nationalities
Wuhan, China

Verena M. Dirsch
Department of Pharmaceutical Sciences
University of Vienna
Vienna, Wien, Austria

ISSN 2191-7043 ISSN 2192-4309 (electronic)
Progress in the Chemistry of Organic Natural Products
ISBN 978-3-030-89872-4 ISBN 978-3-030-89873-1 (eBook)
https://doi.org/10.1007/978-3-030-89873-1

This Springer imprint is published by the registered company Springer Nature Switzerland AG
The registered company address is: Gewerbestrasse 11, 6330 Cham, Switzerland

Preface

Over the last 200 years, higher plants have continued to afford numerous important single-chemical entity drugs for a variety of human illnesses, including colchicine, digoxin, galanthamine, morphine, paclitaxel, and vincristine. Moreover, it is perhaps surprising that even in the last five years, compounds of plant origin are still gaining drug approval by the U.S. Food and Drug Administration. Examples include cannabidiol (for the treatment of Lennox-Gastaut and Dravet syndromes), cocaine hydrochloride (for use as a local anesthetic in nasal surgery), and D-mannitol (for adjunct therapy of adult patients with cystic fibrosis).

As detailed in this highly insightful volume by Professors David Kingston and Maria Cassera, the role of plant compounds in producing antimalarial chemotherapeutic agents has been especially meritorious. The two lead compounds quinine and artemisinin have been extremely important in this regard, although at different times in history. The alkaloid quinine was isolated initially in pure form from *Cinchona* bark by the French investigators Pelletier and Caventou in 1820. As detailed herein, quinine and its key semi-synthetic derivatives such as chloroquine and primacrine have had very important roles in treating malaria, which was rampant even in some temperate countries of the world until quite recently. Artemisinin (qinghaosu), was discovered from a plant used in Chinese traditional medicine about 50 years ago, and, for this breakthrough, Professor Youyou Tu shared the Nobel Prize in Physiology or Medicine in 2015. Artemisinin is a sesquiterpene endoperoxide, and a number of analogs are now utilized in combination chemotherapy for malarial infections, including artemether, arteether, and artesunate. The last-mentioned derivative was approved by the U.S. FDA in 2020, for administration as an intravenous injection to treat severe malaria in adults and pediatric patients. Unfortunately, *Plasmodium falciparum*, the main causative organism of many malarial infections, has become resistant to quinine and its derivatives, and also is showing evidence of resistance to analogs of artemisinin. Therefore, the search for new antimalarial agents of natural origin is continuing, and approaches to their discovery are surveyed in the present contribution. Toward the end of this volume, a wide range of secondary metabolite structural types with demonstrated antimalarial activity, sourced not only from higher plants but also from terrestrial microbial and marine organisms, are discussed for their promise for further development.

The interest of the present authors in the search for new antimalarial agents emanates from their involvement in an International Cooperative Biodiversity Groups (ICBG) project funded by the Fogarty Center, National Institutes of Health, Bethesda, MD, USA, which was focused, in part, on investigating the bioactive constituents of selected species of the flora of Madagascar. Later on, they carried out a project to screen the plant extract collection prepared originally by Merck & Co. Pharmaceutical Company, which is housed currently at the Baruch S. Blumberg Institute, Doylestown, PA, USA. The types of compounds shown with antiplasmodial activity by Professors Kingston and Cassera and their associates from these projects have been representative of a range of plant natural product structural types, inclusive of anthraquinone, indolizinium alkaloid, phloroglucinol, and quinolone alkaloid derivatives.

Columbus, OH, USA — A. Douglas Kinghorn
Linz, Austria — Heinz Falk
Norwich, UK — Simon Gibbons
Tokushima, Japan — Yoshinori Asakawa
Wuhan, China — Ji-Kai Liu
Vienna, Austria — Verena M. Dirsch

Contents

Contributors

Maria Belen Cassera Department of Biochemistry and Molecular Biology, and Center for Tropical and Emerging Global Diseases (CTEGD), University of Georgia, Athens, GA, USA

David G. I. Kingston Department of Chemistry and the Virginia Tech Center for Drug Discovery, Virginia Tech, Blacksburg, VA, USA

Antimalarial Natural Products

David G. I. Kingston and Maria Belen Cassera

Contents

The original version of this chapter was revised (an incorrect series of formulas in the formula scheme on Page 70 has been updated with the correct formula). The correction to this chapter can be found at
https://doi.org/10.1007/978-3-030-89873-1_2

D. G. I. Kingston (✉)
Department of Chemistry and the Virginia Tech Center for Drug Discovery, Virginia Tech, Blacksburg, VA 24061, USA
e-mail: dkingston@vt.edu

M. B. Cassera
Department of Biochemistry and Molecular Biology, and Center for Tropical and Emerging Global Diseases (CTEGD), University of Georgia, Athens, GA 30602, USA
e-mail: Maria.Cassera@uga.edu

A. D. Kinghorn, H. Falk, S. Gibbons, Y. Asakawa, J.-K. Liu, V. M. Dirsch (eds.), *Antimalarial Natural Products*, Progress in the Chemistry of Organic Natural Products 117, https://doi.org/10.1007/978-3-030-89873-1_1

1 Introduction and History of Malaria

Malaria! The very word conjures up visions of sickness and disease, and well it might. The disease has been known since ancient times, with cuneiform tablets from Mesopotamia recording malaria-like deadly fevers, and references to tertian and quartan fevers typical of malaria in the 270 BC Chinese medical text "Nei Chin" [1]. Homer mentions malarial fevers in The Iliad "Sirius, harbinger of fevers, the evil star which dominates the night sky at harvest time" [2], as do several other Greek writers including Hippocrates [2]. It has been asserted that malaria was a major cause of the decline of the Greek City-States [3], although this conclusion has been debated [4]. The most dangerous form of the disease has been shown to be present in first–second century CE in southern Italy [5], so it may have been one of the causes of the decline and fall of the Roman Empire. It was originally thought by the ancient Greeks and Romans to be caused by bad-smelling air, since it has a high incidence in hot and humid low-lying swampy areas with foul-smelling air, and these areas were thus described as having "bad air", or "mala aria" in Italian. Many of these areas were in Italy, particularly around the Vatican in Rome, and this caused real problems for the papacy. Several popes died of malaria, including Alexander IV in 1503, Leo X in 1521, Sixtus V in 1590, and his successor, Pope Urban VII, in the same year after only 13 days as Pope, and papal conclaves were fraught with danger for visiting cardinals [6]. The word malaria itself was introduced into the English scientific literature by John Macculloch in place of the names swamp fever or marsh fever or ague that were then commonly used [7].

Malaria remained a problem in much of Europe and the USA until the early twentieth century. In addition to its severe effects in Italy, it was widespread in the rest of Europe, especially including the low-lying areas of the Netherlands and eastern England, and its existence in Anglo-Saxon England has been inferred from the prevalence of the skeletal condition cribra orbitalia [8]. It was only eradicated in England in the late nineteenth century by a combination of swamp drainage and increased cattle density [9]. Malaria in its milder *Plasmodium vivax* and *P. malariae* forms most probably arrived in the USA with European settlers, but the more dangerous *P. falciparum* variety came with African slaves, who were partly resistant to the disease because of the prevalence of sickle-cell anemia in sub-Saharan Africa [1]. Sadly, it was this very resistance of Africans to malaria that made them sought after as slaves in the Southern USA, and was thus partly responsible for slavery and the horrible slave trade.

Malaria has shaped history in many ways. In one fascinating insight, Charles Mann suggests that malaria helped bring about the union of Scotland and England to form Great Britain. In the late 1600s, Scottish merchants established a trading colony in Panama on a joint stock basis, but the colony failed, in part due to the prevalence of malaria in Panama, and many people lost all their money. England agreed to pay the debts of the stockholders in full in return for unification of the two countries, which was achieved by the Union Act of 1707 [10]. In the USA, the disease devastated both Union and Confederate armies in the American Civil War in the mid-nineteenth

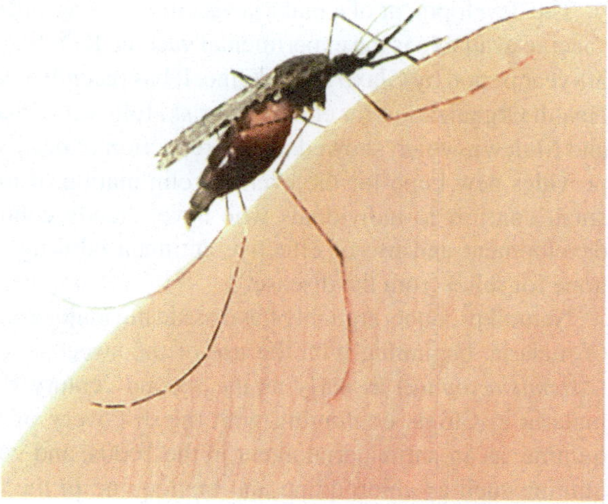

Plate 1 *Anopheles* mosquito; photograph courtesy Dr. Chester J. Joyner, University of Georgia, USA

century and was a major factor in the relatively slower economic development of the southern states. On the other hand, discovery of the malaria parasite by Alphonse Laveran in the 1880s and of the *Anopheles* mosquito (Plate 1) as the malaria vector by Ronald Ross in 1897 [11] opened up ways to reduce the incidence of malaria by controlling the mosquito population, and the building of the Panama Canal was greatly facilitated by measures taken to protect against mosquitos [12]. Even so, there were as many as 500,000 cases of the disease in the USA as late as 1914, primarily in the southeastern states, and malaria was not finally eradicated in the USA until 1951 after the launch of the U.S. National Malaria Eradication Program in 1947 [1].

In terms of human suffering, the greatest impact of malaria today is in sub-Saharan Africa and in Southeast Asia, with an estimated 229 million cases and 409,000 deaths attributed to it in 2019, and with 67% of these deaths occurring in children under 5 years of age [13]. In spite of this huge and devastating impact, research on the discovery and development of antimalarial drugs has been largely neglected by scientists in the developed world, although this is changing as the impact of climate change makes possible a wider spread of the disease [14]. One very encouraging development is the recent announcment of the WHO that China is now officially malaria-free [15].

Reduction of the malaria burden can in principle be achieved by reduction of the mosquito population or of its access to human beings, by development of a malaria vaccine, or by drug treatment of affected individuals. Reduction of the mosquito population proved effective in the years up to 1960, and malaria was eliminated from Cyprus in 1950 by an *Anopheles* eradication campaign that made extensive use of DDT (dichlorodiphenyltrichloroethane) [16]. Unfortunately, the persistence of DDT in the environment led to its widespread banning, and mosquito populations and the incidence of malaria rebounded. Nevertheless, a combination of indoor spraying, bed nets, and new medicines has led to the elimination of malaria in Sri Lanka [17] and reduction of its incidence in Africa and Southeast Asia.

The development of a malaria vaccine is a very difficult task, and to date, the only vaccine available is the experimental vaccine RTS,S/AS01 that was developed over a 30-year period by Glaxo SmithKline. It has recently been recommended by the World Health Organization for children at risk, following Phase 4 studies in Ghana, Kenya, and Malawi, which showed a 30% reduction in deadly severe malaria. This vaccine provides new hope for the eventual elimination of malaria [18]. This is, however, small comfort to individuals who have already contracted malaria, for whom the development and use of effective antimalarial drugs is the only option that offers hope for relief from the disease.

Natural products arguably have made the major contribution to the drug treatment of malaria, beginning with the use of the sweet sagewort plant named "Qinghao" (*Artemisia annua*) as early as the second century BCE [19] for the treatment of malaria in China, continuing with the discovery of *Cinchona* bark, the source of quinine, as an antimalarial agent in the 1600s, and with the synthesis of analogs of quinine such as chloroquine and mefloquine in the 1940–1960s, and culminating in the discovery of artemisinin in the 1970s [20]. This review will focus on the continuing role played by natural products in the treatment of malaria, including a brief introduction to the complex biology of malaria, the historical importance of quinine and artemisinin and other clinically used natural products, and the discovery of potential new natural product drugs.

2 Malaria Biology

2.1 The Malaria Parasite

Malaria is a systemic disease caused by protozoan parasites of the phylum Apicomplexa, class Aconoidasida, order Haemosporida, family Plasmodiidae, and genus *Plasmodium.* There are more than 100 species of *Plasmodium* that infect many animal species including reptiles, birds, and mammals. Presently, five species have been identified that infect humans. *Plasmodium falciparum* (Plate 2) was originally described by William Welch in 1897 and causes malignant tertian malaria, characterized by severe or fatal disease with periodic febrile responses every 36 or 48 h. Independently of the species, the periodic febrile response (paroxysm of malaria) is caused by rupture of mature schizonts. *Plasmodium vivax* was originally described by Giovanni Battista Grassi and Raimondo Feletti in 1890 and causes the tertian benign malaria as the periodic febrile response occurs every 48 h. *Plasmodium falciparum* and *P. vivax* account for the majority of cases of malaria worldwide. *Plasmodium ovale* was originally documented by Stephens in 1922 and also causes a tertian benign malaria. Infections with *P. ovale* are widely distributed but rarely investigated, and recently two subspecies, *P. ovale curtisi* and *P. ovale wallikeri,* were described [21]. *Plasmodium malariae* was the first *Plasmodium* species to be discovered in the

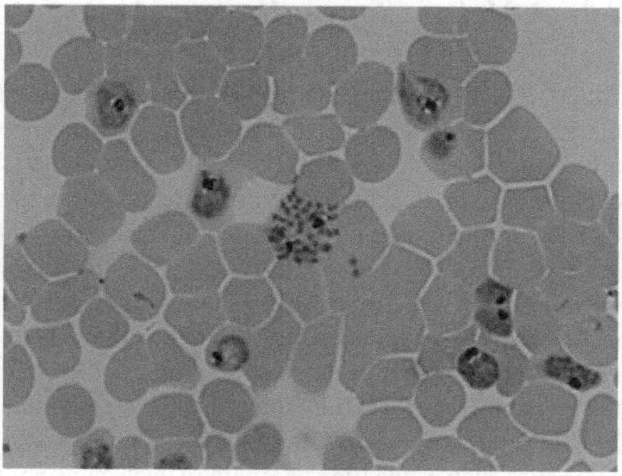

Plate 2 *Plasmodium falciparum*, asexual blood stages

blood of malaria patients by Alphonse Laveran in 1881 when our understanding of the malaria parasites actually began. In *P. malariae* disease, fever occurs every 72 h causing quartan benign malaria. For several decades, these were the only four species described to infect humans, and, with the exception of *P. malariae* that also infects non-human primates, the other three can only infect humans. This has supported the hypothesis that they evolved with our species. *Plasmodium knowlesi* has a 24-h blood stage cycle and mainly infects non-human primates. Although this species was described as early as the 1920s, it was not until the 1960s that occasional human malaria caused by this species was confirmed. Today, *P. knowlesi* is recognized as the fifth species that causes human malaria. A more detailed history about human malaria species is summarized and reviewed by Francis Cox [22].

Each species has its own characteristics, and several reviews can be found describing their individual features and peculiarities [23–27]. Some of their peculiarities worth mentioning are differences in the blood host cell preference, as these have an impact on drug discovery and development of therapeutics. For example, while *P. falciparum* infects mainly mature erythrocytes, *P. vivax* has a strong preference for immature reticulocytes, which offer a richer metabolic environment to the parasite than erythrocytes [28]. Another important characteristic is the dormant liver stage of *P. vivax* called hypnozoites (from Greek "Υπνος", *hypnos*, sleep) that can remain in that state for weeks, months or even years after the primary mosquito-transmitted infection, thus causing malaria relapse episodes [23, 29]. The dormancy phenomenon has also been suggested to occur in *P. ovale,* although hypnozoites of *P. ovale* have never been documented experimentally [30]. *Plasmodium falciparum* is the most studied of all the species, not only because it is the deadliest of the *Plasmodium* species but also because it was the first species where continuous in vitro culture was successfully established in a laboratory setting by Trager and Jensen in the 1970s [31], and it remains the only widely cultured species in laboratories around the world.

2.2 A Malaria Tango: Host and Vector

Tango! The first phrase that comes to mind is that "it takes two to tango". This is the case with malaria, since the malaria parasite needs a vertebrate host, such as a human, and an invertebrate host (the mosquito vector) to maintain its perpetual cycle. However, the comparison goes deeper. Both the dance and lyrics of tango are full of passion, complicated circles of love, loss, and disappointment, which very much resemble the complex life cycle of the malaria parasite between the two hosts. Moreover, the tango rhythm is originally written as a two by four ("dos por cuatro") resembling the "two" hosts and the "four" obligatory cycles or developments: the exo-erythrocytic cycle (liver-stage development), the asexual erythrocytic cycle (blood), the gametocytogenesis (blood/bone marrow), and the sporogonic cycle (mosquito). This general life cycle is similar between species, but unique species-specific characteristics such as the time or length of each cycle are important features to take into consideration in the development of effective antimalarials. In this contribution, we will focus on *P. falciparum*, for which the life cycle is illustrated in Fig. 1.

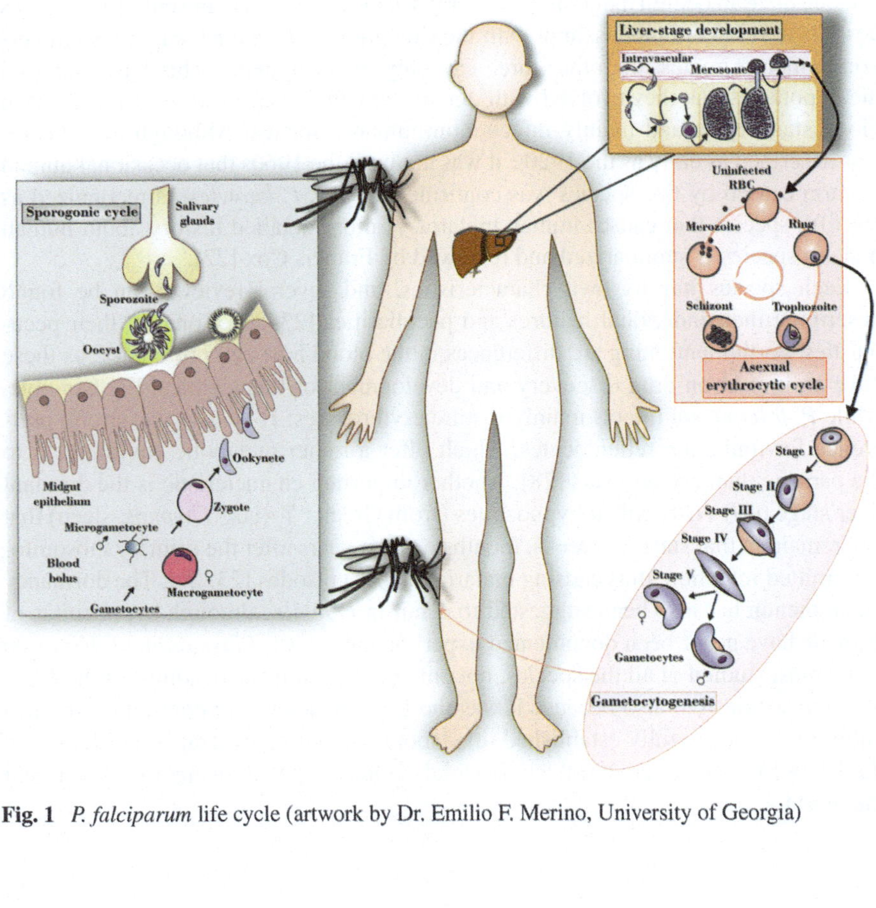

Fig. 1 *P. falciparum* life cycle (artwork by Dr. Emilio F. Merino, University of Georgia)

2.2.1 *P. falciparum* Life Cycle in the Human Host

The parasitic life cycle starts when the female mosquito bites a human during a blood meal and sporozoites that have accumulated in the salivary gland are inoculated into the host along with saliva. Within 1 to 3 hours, sporozoites travel to the liver and the exo-erythrocytic cycle (asexual replication or schizogony) begins and lasts around 10 days. During that time, sporozoites form a parasitophorous vacuole and differentiate into schizonts. The schizogony process involves thousands of mitotic replications generating merozoites that are released into the blood stream and invade mature erythrocytes. The intraerythrocytic schizogony or asexual erythrocytic life cycle is repeated in regular cycles that last around 48 h, during which time the parasite progresses through four morphologically different stages named ring, trophozoite, and schizont, ending with the rupture of the erythrocyte and the release of merozoites that will invade new erythrocytes. The asexual erythrocytic life cycle is responsible for the manifestation of signs and symptoms in malaria [27]. The majority of the intraerythrocytic schizogony occurs in the capillary beds of organs, and schizonts are rarely found in the peripheral blood due to their capability to cytoadhere to the endothelial cells [32].

Transmission of the malaria parasite requires the development of different parasite forms, called male and female gametocytes, non-replicating forms that can persist for weeks in circulating blood until they are ingested by female mosquitoes during a blood meal. Gametocytes develop during intraerythrocytic stages when an asexual schizont is committed to sexual reproduction. The rate of commitment to sexual development varies widely between species, and in recent years, considerable progress has been made in elucidating the molecular mechanisms governing this important differentiation process [33–36]. *Plasmodium falciparum* gametocytes take between 10 and 12 days to mature and progress through five morphologically distinct forms (stages I–V), which are different from other *Plasmodium* species. Only the most mature forms of *P. falciparum* gametocytes are present in the circulation, whereas immature forms accumulate in the hematopoietic environment of the bone marrow [36, 37]. While gametocytes inflict no clinical symptoms in humans, they are responsible for contributing to the transmission cycle of malaria [38]. Therefore, without a highly effective vaccine [39], combinations of malaria prevention and treatment are required, and the gametocyte stages of the parasite's life cycle represents an ideal target for novel strategies to block transmission [38].

2.2.2 *P. falciparum* Life Cycle in the Mosquito Vector

After a female anopheline mosquito ingests gametocytes during a blood meal, sexual reproduction occurs in the midgut [40]. During the blood meal, gametocytes experience a dramatic change in environment and must adapt adequately. A decrease in temperature from 37°C to ambient temperature and the presence of xanthurenic acid in the mosquito midgut activate the gametocytes to undergo gametogenesis, in which both male (microgametocyte) and female (macrogametocyte) forms egress from infected erythrocytes [41]. The mature microgametocyte rapidly replicates its genome three times, progressing from haploid to octaploid, and mitotically produces eight flagellar microgametes, a process termed exflagellation. After detaching from the residual body, a microgamete is freely motile in search of a macrogamete. On the other hand, the non-motile macrogametes contain a well-developed endoplasmic reticulum, mitochondrion, and apicoplast in preparation for rapid development as a zygote [42, 43]. The formation of the zygote is usually complete within 30 min [42, 44–47]. The formed zygote secretes a chitinase along with other micronemal secretory proteins that allows the parasite to penetrate the peritrophic membrane surrounding the blood meal [48, 49]. The ookinete will then migrate through the midgut epithelium to the basal lamina, where it will develop into an oocyst [48, 50]. The oocyte generates sporozoites that travel to the salivary glands of the mosquito and will be injected into another human host upon the mosquito's next blood meal.

3 Natural Products as Antimalarial Agents

3.1 *Quinine*

3.1.1 Discovery and History

As noted in the introduction, natural products have made a huge impact on malaria treatment, and quinine in particular legitimately can be named as a drug that changed the world. Before the discovery of the beneficial properties of *Cinchona* bark, the source of quinine, treatment of malarial fevers in Europe involved everything from crab's eyes to cat's blood and sage leaves, although bleeding and purgation were the most common treatments [51]. These treatments were uniformly ineffective, and in most cases, did more harm than good.

The details of the discovery of the curative effects of *Cinchona* bark (Plate 3) on malaria are unknown, but the legend that the Countess of Cinchón, wife of the Spanish Viceroy of Peru, was cured of the ague by a native remedy consisting of the bark of a tree is almost certainly untrue [51]. It is also reasonably certain that malaria did not exist in the New World until it was brought by European colonists, so it is unlikely that the Peruvian Indians discovered *Cinchona* bark as the cure for this non-existent disease. An intriguing theory of the discovery of the value of *Cinchona* bark is that the Peruvian Indians drank hot infusions of *Cinchona* bark to suppress shivering when they were cold, and that the Spanish colonists tried the same infusions to counteract the shivering of malaria [51–53], and miraculously found that these actually cured the disease. Be this as it may, the value of infusions of *Cinchona* bark as a cure for malaria was recognized, and the botanist Carl Linnaeus later named the tree that is the source of quinine "Cinchona" in honor of the Countess.

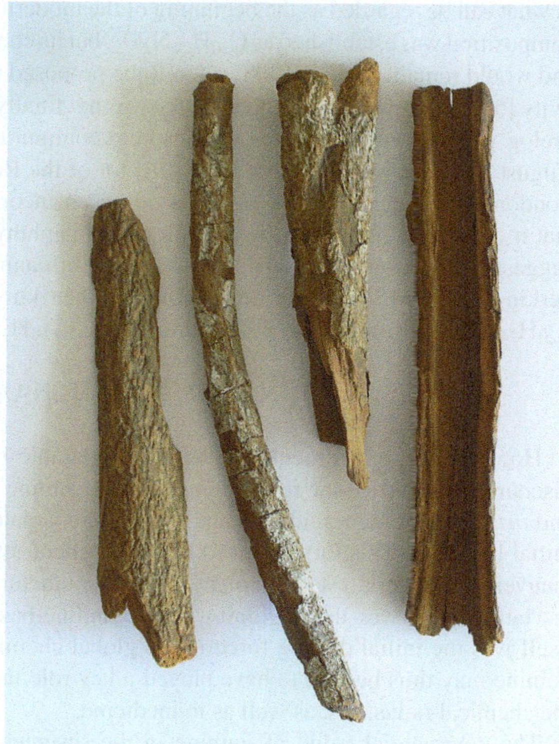

Plate 3 *Cinchona* bark—*Cinchona officinalis*; photgraph courtesy H. Zell, licensed under Creative Commons Attribution

In the 1640s, Father Bartolomé Tafur took some bark back to Rome where it was embraced by the clergy, with the result that the papal conclave of 1655 was the first without one of the attending cardinals dying from malaria [54]. Recognizing the medical and commercial value of *Cinchona* bark for the treatment of malaria, the Jesuit missionaries who were part of the Spanish colonial expeditions began the large-scale export and sale of the bark to Europe, where it became known as "Jesuits' bark". The cure was not valued in Protestant England, and Oliver Cromwell, the Lord Protector of England during the Commonwealth, is thought to have died in 1658 from a combination of malaria and typhoid fever [55]. To overcome this problem, Robert Talbor, a London apothecary, developed a secret "English Remedy…for Cureing of Agues and Feavers," and warned of the dangers of using Jesuits' bark. His cure was used successfully by Cromwell's successor, King Charles II, and Talbor was awarded £20,000 (an enormous sum at the time) for this cure. After his death, the "secret remedy" turned out to be Jesuits' bark and wine.

The active component of *Cinchona* bark was first isolated in pure form in 1820 and named quinine by the French chemists Pierre-Joseph Pelletier and Joseph Caventou, and by 1826 these chemists were producing over 3500 kg of quinine sulfate each year in what can be regarded as the beginning of the modern pharmaceutical industry.[1] Its composition was established as $C_{20}H_{24}N_2O_2$, but its chemical structure was unknown and would remain so until 1908, when Rabe proposed the correct structural connectivity [55]. Its complete stereochemistry was not finally established until 1944 when Prelog worked out its relative and absolute configurations [56]. This did not stop August Wilhelm von Hofmann, then director of the Royal College of Chemistry in London, from suggesting in 1849, based on an incorrect composition of quinine, that it could be synthesized by oxidation of α-naphthylamine. It is not clear if this suggestion was ever followed up, but in 1856 Hofmann's star student William Henry Perkin attempted to synthesize quinine, by then known to have the composition $C_{20}H_{24}N_2O_2$, by oxidation of *N*-allyl toluidine ($C_{10}H_{13}N$):

$$C_{10}H_{13}N + 3\,O \longrightarrow C_{20}H_{24}N_2O_2 + H_2O$$

However, the experiment yielded an inseparable black tar, but Perkin was not discouraged and tried the reaction on a simpler amine (actually a mixture of aniline and *ortho*- and *meta*-toluidines) and was able to isolate a purple substance from the initial black tar. This turned out to be an excellent dye that was eventually named mauveine, and Perkin set up a plant for its manufacture with financial support from his father. This was the beginning of the aniline-based dyestuffs industry, which itself was the initial driving force in the global chemical industry [57]. In a sense, quinine may thus be said to have played a key role in both the pharmaceutical and the chemical industries, as well as in medicine.

The commercial value of quinine in the seventeenth and eighteenth centuries made the cultivation of *Cinchona* trees and the export of their bark a boon for the economies of Ecuador, Peru, and Bolivia, and the export of *Cinchona* seeds was strictly forbidden. This did not stop Europeans from trying to obtain seeds, and both British and French explorers obtained seeds of *Cinchona calisaya* (Plate 4) and *Cinchona succirubra.* Although *C. calisaya* did not take well in foreign soils, plantations of the latter species were established in India, Ceylon, and Java. Several pounds of seeds of a high-producing strain of *C. calisaya* were later collected with great difficulty by the Englishman Charles Ledger, who sent them to England. The English government was not interested in purchasing them, but the Dutch government was, and used these seeds to establish plantations in Java using grafted varieties of the plant [51]. These plantations produced over two-thirds of the world supply of quinine until World War II, when the Japanese control of Java forced the Allies to seek alternative supplies and synthetic substitutes for quinine.

Quinine continues to be a significant commercial product, although its major use today is in the preparation of tonic water. This dates back to the British Raj in India, when the colonizers would take regular doses of quinine in water as a prophylactic

[1] Note added in proof: The authors recently became aware of the article, Anft B (1955) "Friedlieb Ferdinand Runge: a forgotten chemist of the nineteenth century", J Chem Ed 32:566 stating that Runge reported the isolation of quinine, which he called "China base", in 1819, a year before the French team.

Plate 4 Branches of *Cinchona calisaya* (Ledger quinine), Makawao Forest Reserve, Maui, Hawaii; photograph courtesy Forest & Kim Starr, licensed under Creative Commons Attribution

against malaria. Since quinine has a bitter taste, the drug was diluted with gin to mask the taste, giving rise to gin and tonic. The tonic water used today contains much less quinine than the prophylactic dose. Although it is no longer used prophylactically, quinine also continues to be used for the treatment of some drug-resistant malarias.

3.1.2 Structure and Biosynthesis

The structure and stereochemistry of quinine, as deduced by Rabe and Prelog, is (*R*)-[(2*S*,4*S*,5*R*)-5-ethenyl-1-azabicyclo[2.2.2]octan-2-yl]-(6-methoxyquinolin-4-yl)methanol (**1**) using IUPAC nomenclature, which differs from Rabe's original numbering. *Cinchona* bark contains quinine and a mixture of related alkaloids, including cinchonidine (**2**), quinidine (**3**), and cinchonine (**4**).

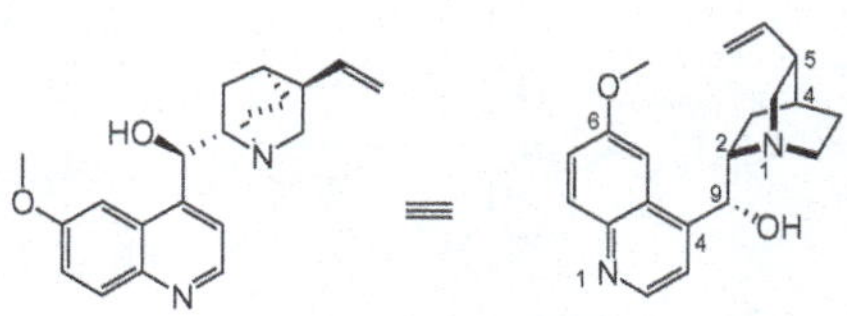

1 (quinine)

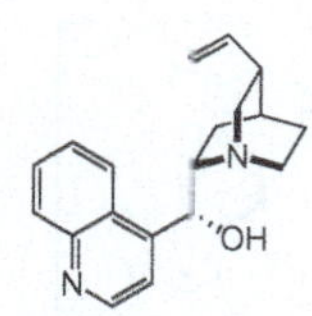

2 (cinchonidine)

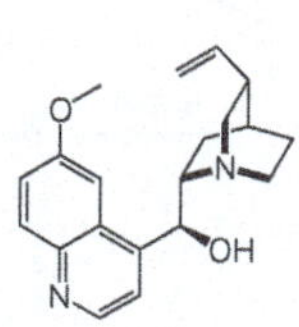

3 (quinidine)

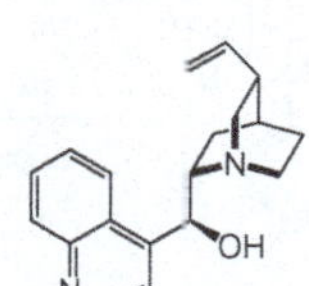

4 (cinchonine)

The four major *Cinchona* alkaloids differ only in the presence or absence of a methoxy group at C-6 and in the configuration of the hydroxy group at C-9. Cinchonidine (**2**) and cinchonine (**4**) both have antimalarial activity, although they are not as effective as quinine (**1**). Quinidine (**3**) is used to treat cardiac arrhythmias, and although undesired side effects have limited its application, recent reviews support its continued use in modern cardiology [58, 59].

The biosynthesis of quinine (**1**) was first studied in detail by Leete [60] and later by Battersby [61–63], and the overall pathway is summarized in Scheme 1 [64]. Labeling and precursor studies gave the surprising result that the biosynthesis starts with tryptophan, and thence to the known indole alkaloid strictosidine (**5**); strictosidine synthase has been purified from a suspension culture of *Cinchona robusta* [65]. Conversion of **5** by established pathways gives the known indole alkaloid corynantheal (**6**), which is where the pathway diverges from those of other indole alkaloids. Oxidation of **6** at C-5 and imine formation between the resulting amine and the aldehyde group followed by reduction yields cinchoaminal (**7**), which then undergoes C–N cleavage to yield the aminoketoaldehyde **8**. Cyclization of **8** yields cinchonidinone (**9**), which is reduced to cinchonidine (**2**) or reduced, hydroxylated, and methylated to give quinine (**1**). Three enzymes involved in the early and late steps of the pathway were recently identified by analysis of transcriptomic data from *Cinchona pubescens* [66]. So quinine yet again surprises as an "indole alkaloid" without an indole ring.

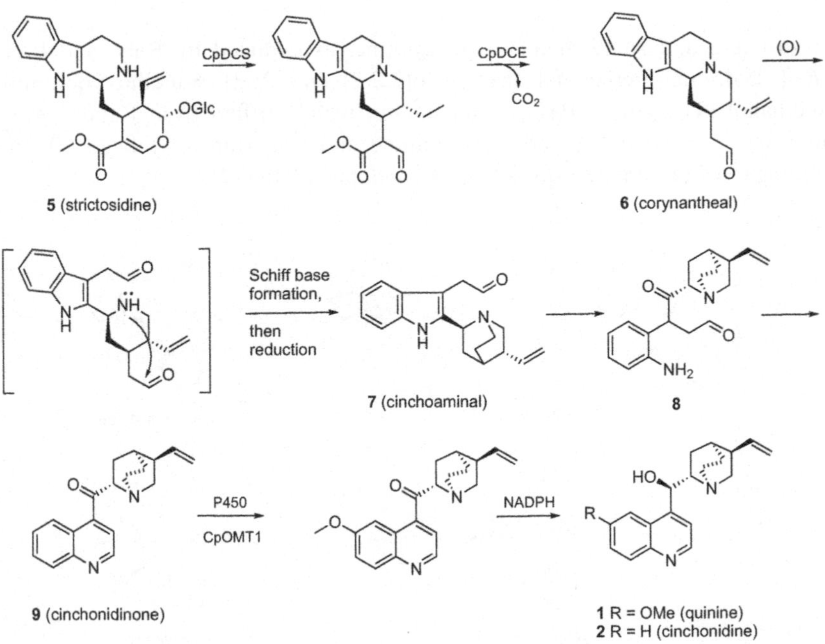

Scheme 1 Biosynthesis pathway of quinine (**1**) from strictosidine (**5**)

3.1.3 Mechanisms of Action and Resistance

As described in detail in Sect. 2, the malaria parasite enters the human host as a sporozoite after a bite from an infected mosquito. The sporozoites then enter the blood stream and move to the liver, where they invade the liver hepatocytes and mature over a period of a week, multiplying asexually, and then rupture the host cells and reenter the bloodstream as merozoites, where they invade red blood cells. In order to survive and multiply, the parasites digest host hemoglobin as their food source, aided by the aspartic proteinase enzymes plasmepsins I and II and the cysteine protease falcipain. This process releases heme into the food vacuole, which undergoes autooxidation to hematin and the formation of hydrogen peroxide. Since hematin is toxic to the parasite, it is detoxified by conversion to its insoluble polymeric form hemozoin.

The key to quinine's antimalarial activity is its ability to inhibit the formation of hemozoin, thus dooming the merozoites to death by their toxic hematin product. The exact way that quinine does this was not known until recently, when de Villiers et al. published the single crystal X-ray diffraction structures of the complexes formed between quinine (**1**) and quinidine (**3**) and iron(III) protoporphyrin IX (Fe(III)PPIX) [67]. These structures showed that both quinine (**1**) and quinidine (**3**) form five-coordinate complexes with the porphyrin through their benzylic C-9 hydroxy groups, as well as an intramolecular hydrogen bond between the propionate side chain of Fe(III)PPIX and the protonated quinuclidine nitrogen atom of the alkaloid. This complex then inhibits the formation of hemozoin. This finding is fully consistent with earlier synthetic work that showed that a hydroxy functionality and a rigid proximal N are required for forming a non-covalent quinine–heme complex [68]. A deeper understanding of this mechanism was added by Rimer and Vekilov, who showed that quinine (**1**) and related antimalarials inhibit hematin crystallization by three distinct drug–surface site interactions [69].

Although quinine (**1**) has been in use in one form or another since the mid-seventeenth century, resistance to it was slow to develop and was only reported for the first time at the beginning of the twentieth century [70]. Even today resistance is usually low, so **1** is still efficacious in most cases [71]. The mechanism of resistance is however still unknown [72], although resistance may be associated with single nucleotide polymorphisms (SNPs) in pfmdr1, pfmdr6, pfcrt, pfmrp1, and pfnhe1 [69, 73].

3.1.4 Clinical Use

While quinine (**1**) is still an important antimalarial drug, it has a low therapeutic index and significant side effects, including reversible hearing impairment, headaches, abdominal pain, vomiting, and diarrhea [71]. Its use as a first-line treatment is thus limited, but it is still valuable as an adjunct to other treatments, and can be life-saving. In one recent example, two patients who did not respond to treatment with parenteral artesunate for severe malaria were then treated with **1**. Sadly, one patient died, but the other survived although he did not clear his peripheral parasitemia until 11 days later [74]. Achan et al. concluded their review of the role of **1** in malaria therapy by stating "In the near future, quinine will continue to play a significant role in the management of malaria, particularly in resource limited settings....Quinine continues to play a critical role in the management of malaria in the first trimester of pregnancy, and will remain so until safer alternatives become available. The continued use of quinine in the management of uncomplicated malaria is a concern. Clearly, the seven-day duration of therapy and three times daily administration of quinine present a major challenge to the completion of therapy, leading to sub-optimal treatment outcomes. In these situations, artemisinin-based combination therapy (ACT) is a better option given the simplicity of dosing and shorter treatment duration" [71].

3.2 Quinine Analogs

The side effects of the use of quinine (**1**) as an antimalarial agent as well as its cost prompted chemical and pharmaceutical companies to search for alternative drugs. Although these compounds are not natural products, they were synthesized as analogs of **1**, and are thus included in this contribution as examples of the crucial role of quinine (**1**) in antimalarial drug discovery. This search for replacements for **1** was initiated by the German chemical industry in the 1920s following the loss of access by Germany to the world's major quinine supplies at the end of World War I, and was driven both by the price of **1** and by its perceived shortcomings [75]. This work resulted in the syntheses of pamaquine (**10**), quinacrine (**12**), also known as mepacrine and by its tradename Atabrine, sontaquine (**13**), and chloroquine (**14**). The loss of access by the Allies to quinine supplies in World War II following the German conquest of the Netherlands and the subsequent occupation of the then Dutch East Indies by the Japanese in March 1942 was huge, since these islands provided 90% of the world's quinine supplies. This was especially significant as much of the war would need to be fought in malarial areas of SE Asia and southern Europe. Ironically, the compound that ended up saving the lives of many of the Allied forces in World War II was Atabrine (**12**), first synthesized by German scientists in the 1930s.

Eight such compounds, pamaquine (**10**), primaquine (**11**), quinacrine or Atabrine (**12**), sontaquine (**13**) chloroquine (**14**), amodiaquine (**15**), mefloquine (**16**), and tafenoquine (**17**) will be discussed briefly in this section (Fig. 2). A detailed review of the mechanisms of action of these and related compounds by Foley and Tilley has been published [76], and it is generally accepted that they work by suppressing hemozoin crystallization by forming non-crystallizable complexes. The synthetic analog of hemozoin is β-hematin, and recent work by Vekilov has shown that quinoline antimalarials have up to three specific interactions with β-hematin crystals, and propose that these drug-hemozoin crystal interactions are key to the antimalarial activity, with complexation with soluble β-hematin playing only a minor role [69, 77].

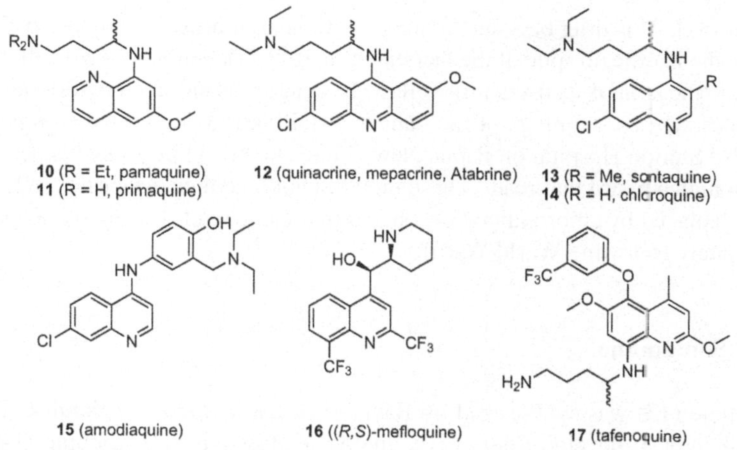

Fig. 2 Structures of selected quinine analogs

3.2.1 Pamaquine

The first effective aminoquinoline drug was pamaquine (**10**), used in 1926 as the first drug with curative activity against *P. vivax* [78]. It was later found to be too toxic for normal clinical use, but it was important as the first example of a synthetic compound with significant antimalarial activity, and paved the way for the discovery of mepacrine.

3.2.2 Primaquine

The toxicity of pamaquine (**10**) was reduced by replacement of the side chain terminal NEt_2 group with an NH_2 group to give primaquine (**11**), which was first synthesized in 1946 in the United States. It has transmission-blocking antimalarial activity and is active against gametocytes of all species of *Plasmodium* parasites. Like many of the aminoquinoline antimalarial agents, primaquine (**11**) has a chiral center on its side chain and is used in its racemic form. In a recent study, the two enantiomers were separated and evaluated individually for their activity, and it was concluded that the (*R*)-configured enantiomer may have a better safety margin than the racemate [79]. A detailed account of its discovery, development, and pharmaceutical properties has appeared [80].

3.2.3 Quinacrine (Mepacrine, Atabrine)

Quinacrine (**12**) was first prepared by Bayer I. G. Farbenindustrie chemists in 1932 as they tried attaching the basic side chain of pamaquine (**10**) to various heterocyclic systems [81]. This drug became the major antimalarial drug used by the Allied war effort in the Pacific, in spite of the fact that its acridine skeleton gave it a yellow color and made the skin of its users turn yellow. Its usage was enforced by providing it at mealtimes, and by warnings of the danger of failing to take it. One sign posted at the 363rd Station Hospital on Papua New Guinea had two human skulls perched on top of a notice board that read "These men did not take their Atabrine" [82]. It was largely replaced by chloroquine (**14**) once this became readily available in the years immediately following World War II.

3.2.4 Sontaquine

Sontaquine (**13**) was synthesized by Bayer I. G. Farbenindustrie chemists in 1936 as an attempt to develop a less toxic analog of Resochin (chloroquine (**14**)), and it performed well in initial evaluations. Information on the drug was shared with the French company Rhone-Poulenc under an existing agreement, and Decourt and Schneider carried out clinical studies on the drug in Tunisia. After the Allied invasion of North Africa on November 9, 1942, Schneider offered his remaining supplies of sontaquine (**13**) to the Allies together with his clinical data. Meanwhile, the Winthrop Chemical Company in New York received the manufacturing directions for the compound under an arrangement with Bayer and arranged for it to be tested for activity. The compound proved to have good activity, but due to confusion over its identity and other factors, its development was delayed until it became clear that chloroquine (**14**) was a superior drug [81].

3.2.5 Chloroquine

The drug now known as chloroquine (**14**) was first synthesized in 1934 by H. Andersag of Bayer I. G. Farbenindustrie and named Resochin. It was evaluated for antimalarial activity against bird malaria and found to be as effective as Atabrine (**12**) but slightly more toxic. It was also tested in four psychiatric patients but was then abandoned by Bayer on the basis of its somewhat greater avian toxicity than **12**; this later became known in Germany as the "Resochin error". The development of chloroquine (**14**) in the USA began in 1943, and it was evaluated in dogs and monkeys, with the latter study showing that it was four times less toxic than sontaquine (**13**). Once patent issues were settled, **14** was evaluated in human trials and then selected as the drug of choice for the management of malaria in 1946 [81, 83].

Chloroquine (**14**) is believed to act by a variety of mechanisms, but one important mechanism is the formation of toxic heme–chloroquine complexes in the food vacuole of the parasite, similar to the proposed mechanism of quinine (**1**) [76, 84].

The availability of a low-cost and less toxic alternative to quinine (**1**) made chloroquine (**14**) the drug of choice for many years, in spite of its low therapeutic index, with an adult curative dose of 10 mg/kg and a toxic dose of 30 mg/kg [76]. Its use was, however, both extensive and indiscriminate; in one example, salt medicated with

14 was provided to over 2,500,000 people in Brazil [85]. Resistance to chloroquine (**14**) developed in the 1960s in Southeast Asia and South America, and resistance has now spread worldwide. Resistance appears to be due to in part to a reduction in the accumulation of **14** in the acidic food vacuole of the parasite as a result of SNPs in pfcrt and pfmdr1 [73], but multidrug resistance due to overexpression of pfmdr1 is probably also involved [76]. A recent study showed that its antimalarial activity is pH-dependent, with a loss of activity as the pH decreases, which suggests that acidosis in malaria patients may be a contributing factor to its decreased activity [86].

3.2.6 Amodiaquine

Amodiaquine (**15**) was developed as a competitor to chloroquine (**14**) by Parke, Davis and Company, but it was later realized that it had led to the death of a number of people because of an unexpected reaction [75]. Its major use is in combination with artesunate to reduce the risk of resistance.

3.2.7 (*R*,*S*)-Mefloquine

Mefloquine (**16**) was discovered as the result of a large-scale screening program at the Walter Reed Army Institute for Research in Washington, DC (USA); from 1963 to 1976 over 250,000 compounds were tested for antimalarial activity. The best compound to emerge from this effort was **16**, and it was marketed as Lariam by F. Hoffmann-La Roche. It is unusual in having two chiral centers, and the marketed drug is a racemic mixture of the (*R*,*S*) (shown) and the (*S*,*R*) enantiomer.

(*R*,*S*)-Mefloquine (**16**) is an effective prophylactic and treatment for malaria and was approved by the U.S. Federal Drug Administration (FDA) in 1989 without the usual randomized Phase III tolerability study. Unfortunately, the drug was later shown to cause neurotoxicity, in some cases so severe that death resulted, and its use now is restricted [87].

3.2.8 Tafenoquine

Tafenoquine (**17**) was approved by the FDA in 2018 as Arakoda™ for prophylaxis of malaria in adults and as Krintafel™ for the radical cure of *P. vivax* malaria in adults [88]. It is an 8-aminoquinoline derivative, unlike chloroquine and most other quinoline antimalarials, which are 4-aminoquinoline derivatives, and is a welcome addition to the arsenal of quinoline antimalarial agents.

3.2.9 Other Quinine Analogs

Various reviews of synthetic antimalarial agents, including other quinine analogs, have appeared over the last 20 years, and are cited here for the interested reader [89–94].

3.3 Artemisinin

3.3.1 Discovery

The sweet sagewort plant "Qinghao", or *Artemisia annua* (Plate 5), was utilized as early as the second century BC for the treatment of malaria in China [19], and its use for treatment of "hot and cold due to intermittent fever illness" was advocated in "Ben Cao Gang Mu" (1596) [95], but these uses were not widely known. In 1964, a

Plate 5 *Artemisia annua*, Botanical Garden Mainz; photograph courtesy Natalie Schmalz, licensed under Creative Commons Attribution

combination of the development of resistance of the malaria parasite to chloroquine (**14**) and Chinese support for North Vietnam in the Vietnam war led the Chinese military to set up a confidential national agency for malaria control known as the 523 office (for the date of May 23 when the project was initiated). Initial work on the 523 project focused on screening thousands of compounds for antimalarial activity, but no effective drugs were identified [20]. In 1969, leaders of the 523 project sought the assistance of the Institutes of Chinese Materia Medica of the Academy of Traditional Chinese Medicine, and Youyou Tu (Nobel Price in Physiology or Medicine 2015 jointly with W.C. Campbell and S. Ōmura) was assigned to lead a search for natural product antimalarial agents. The search began with a review of the literature on Traditional Chinese Medicine and led to the preparation and testing of extracts of over 100 herbs, including "Qinghao", in rodent malaria but with few promising results. The failure of "Qinghao" extracts to show reproducible activity was almost certainly due to the decomposition of artemisinin by the hot water used in the extraction of dried plant material and to the low artemisinin content of this material. The breakthrough came when Youyou Tu went back to one of the original sources, "A Handbook of Prescriptions for Emergencies" by Ge Hong (264–363 A.D.) written in the East Jin Dynasty. In this handbook, Ge Hong recommended soaking the fresh plant in water and then wringing it out, probably producing an emulsion that contained the water-insoluble artemisinin [96]. Youyou Tu and her team eventually worked out an extraction process that used diethyl ether for extraction followed by a basic wash to remove acidic components and found to their delight in October 1971 that this extract was 100% effective in treating rodent malaria. Studies in malarial monkeys confirmed this result in January1972, and clinical trials of the diethyl ether extract later that year on 30 patients were successful, with all patients recovering with full clearance of the *Plasmodium* parasites [20].

The fascinating history of the discovery of artemisinin (**18**) has been reviewed on several occasions in addition to the references cited above [97–102], and these references can be consulted for additional details of the discovery process.

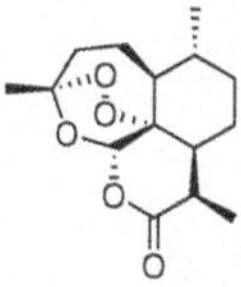

18 (artemisinin)

3.3.2 Structure and Biosynthesis

The initial clinical studies described above were carried out on the partially purified diethyl ether extract of *Artemisia annua*, so the structure of the active compound was not known. It was obtained as a pure compound in late 1972 and named qinghaosu in Chinese and artemisinin in English. Initial analyses indicated its composition as $C_{15}H_{22}O_5$, and its structure was determined by X-ray crystallography as **18** in 1974. Although this result was published in a little-known Chinese journal in 1977 [103] and abstracted by "Chemical Abstracts", the latter publication only showed the structure of artemisinin (**18**) with no mention of its antimalarial activity. The value of **18** as an antimalarial agent thus only became widely known in the Western world with its reisolation by researchers at the Walter Reed Army Institute of Research [104] and by a 1985 review by Klayman [105].

Artemisinin (**18**) is just one of many compounds biosynthesized by *A. annua*; the plant produces almost 600 natural products, with flavonoids and sesquiterpenes predominating, including many terpenoid allylic hydroperoxides. Given this complex phytochemistry, it is a minor miracle that Youyou Tu's ether extract was as effective as it was in treating malaria. The phytochemistry of *A. annua* has been reviewed on several occasions, and will not be discussed here [95, 106–108].

The biosynthesis of artemisinin (**18**) is shown in outline in Scheme 2, and takes place in the glandular trichomes of *A. annua* [110]. It has been shown to proceed from (*E*,*E*)-farnesyl diphosphate to the key intermediate amorpha-4,11-diene (**19**) [110] and thence to artemisinic alcohol (**20**), artemisinic aldehyde (**21**), dihydroartemisinic aldehyde (**22**) [111], dihydroartemisinic acid (**23**), and artemisinin (**18**) [112]. A $^{13}CO_2$ study gave the unexpected result that the central isoprene unit of farnesyl

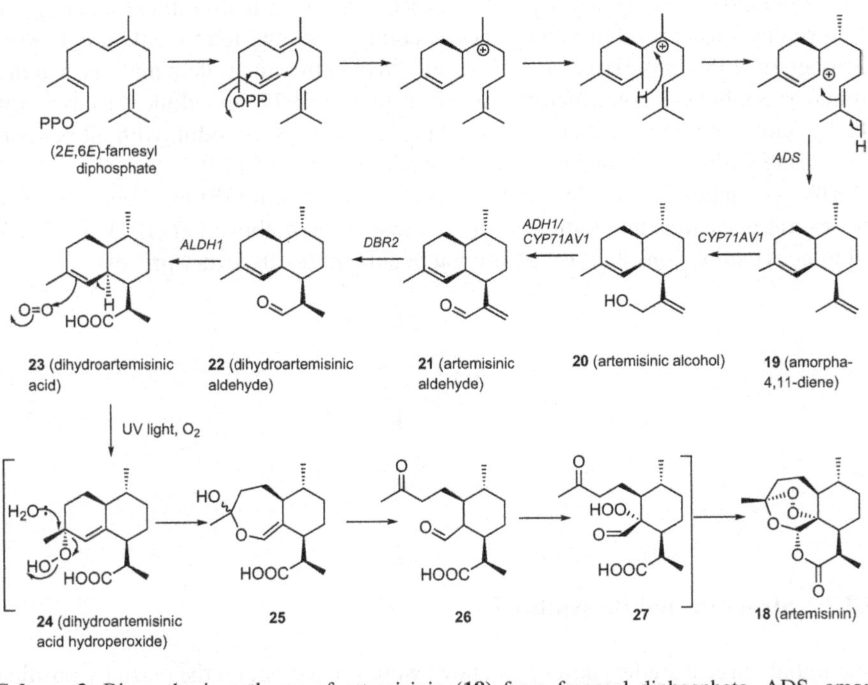

Scheme 2 Biosynthesis pathway of artemisinin (**18**) from farnesyl diphosphate. ADS, amorphadiene synthase; CYP71AV1, amorphadiene monooxygenase; ADH1, alcohol dehydrogenase; DBR2, artemisinic aldehyde, $\Delta^{11(13)}$ reductase; ALDH1, aldehyde dehydrogenase

diphosphate was formed by the non-mevalonate methylerythritol phosphate (MEP) pathway, while the terminal units were both formed by the mevalonate pathway [113]. A later study using [1-^{13}C]acetate, [2-^{13}C]acetate, and [1,6-$^{13}C_2$]glucose confirmed that the MEP pathway contributes to the biosynthesis of artemisinin (**18**) [114]. The conversion of dihydroartemisinic acid (**23**) to **18** has been shown to be nonenzymatic in vivo by feeding studies using ^{13}C-labeled acid [115], and is proposed to follow the pathway shown in Scheme 2. Initial reaction of **23** with oxygen to form dihydroartemisinic acid hydroperoxide (**24**) is followed by rearrangement to hemiacetal **25**, ring opening of **25** to give dialdehyde **26**, a second hydroperoxidation to give **27**, and final formation of two hemiacetals followed by lactonization to yield artemisinin (**18**) [95]. The nonenzymatic nature of this conversion is a benefit to farmers who cultivate *A. annua* for artemisinin (**18**) production, since dried *A. annua* leaves typically contain about 1% each of **18** and dihydroartemisinic acid (**23**), and leaving the leaves to dry in the sun facilitates the conversion of some of the acid to artemisinin (**18**) [116].

3.3.3 Synthesis of Artemisinin and Its Analogs

Artemisinin (**18**) is a moderately complex compound from a synthesis point of view, and is too complex for a total chemical synthesis to be economically viable for the commercial production of what must necessarily be a low-cost drug substance. Nevertheless, several very different synthesis approaches have been successful in producing the desired compound, but in most cases, only after many steps and with low overall yields; reviews and assessments of these strategies have appeared [116–118]. The first synthesis started from (–)-isopulegol, and gave artemisinin in 13 steps and an overall yield of about 5% [119]. A second synthesis by Zhou and coworkers at the Shanghai Institute of Organic Chemistry started with (+)-(*R*)-citronellal and gave artemisinin (**18**) in 0.2% overall yield [120–122]. Formal syntheses of **18** were completed by Ravindranathan et al. [123] and Bhonsle et al. [124], and a Diels–Alder approach by Liu and coworkers appeared in [125]. A synthesis by Avery and colleagues began with (+)-(*R*)-pulegone, and gave **18** in ten steps in about 3.5% yield [126]. Yadav and colleagues prepared artemisinin in 0.5% yield in eleven steps from (+)-isolimonene in [127], and then by an improved route from citronellal in 5% overall yield [128]. The highest yielding synthesis was reported by Zhu and Cook [129] and yielded artemisinin in nine steps and 12–15% overall yield from the relatively inexpensive cyclohexenone (Scheme 3), but even this synthesis is too costly for it to be commercially practical.

A final noteworthy synthesis by Giannis et al. produced *ent*-artemisinin, the enantiomer of the natural product (Scheme 4). Biological evaluation of this synthetic product showed that it had similar antiplasmodial and cytotoxic activity to the natural product **18** [130], suggesting that its mechanism does not involve stereoselective binding to an enzyme or other biological target.

Although artemisinin (**18**) is an effective antimalarial agent, it also suffers from some defects. In particular, it is only sparingly soluble in water, it is not well absorbed

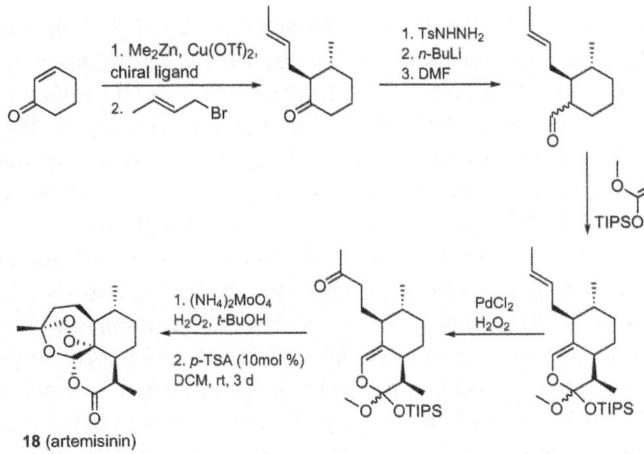

Scheme 3 Synthesis of artemisinin (**18**) from cyclohexenone by Zhu and Cook

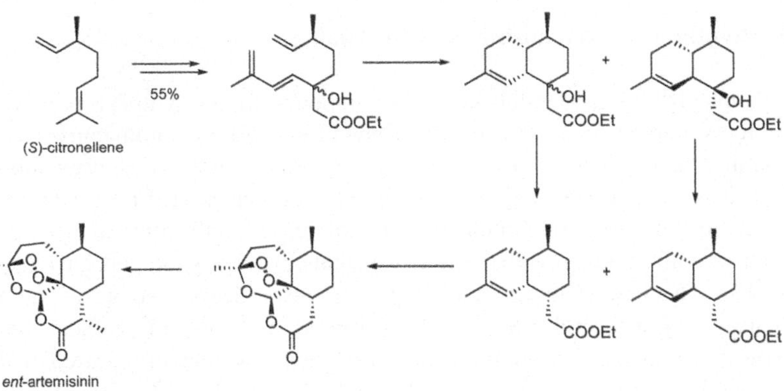

Scheme 4 Synthesis of *ent*-artemisinin by Giannis et al.

from the gastrointestinal tract, and it has a short plasma half-life. For these and other reasons various semisynthetic analogs of artemisinin (**18**) have been prepared, and three of them, dihydroartemisinin, α-artesunate and β-artemether (Scheme 5), are key ingredients of the artemisinin-based combination therapies (ACTs) that are the mainstay of the treatment of *P. falciparum* malaria, as will be discussed later. The half maximal inhibitory concentration (IC_{50}) values for artemisinin and its derivatives vary, most probably due to different assay methods and different strains of *P. falciparum* evaluated. Li et al. reported IC_{50} values of 126, 0.58, and 1.2 n*M* for artemisinin (**18**), dihydroartemisinin, and β-artemether, respectively, against a specific strain of *P. falciparum,* based on incorporation of [G-^{3}H]-hypoxanthine [131]. However, Kumari et al. [132] reported IC_{50} values of 2.4, 2.4, and 1.0 n*M* for these same compounds, but without stating the source of these values.

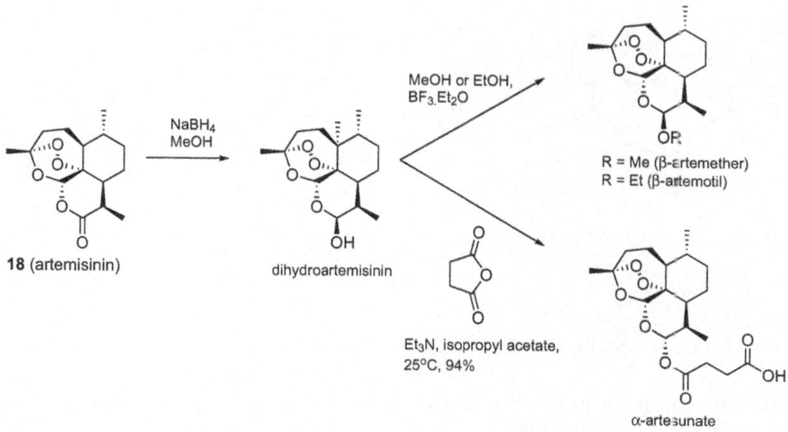

Scheme 5 Conversion of artemisinin (**18**) to dihydroartemisinin, artemether, and artesunate

The conversion of artemisinin (**18**) to its dihydro derivative was originally carried out using potassium borohydride [133], but sodium borohydride works equally well. The reduction of a lactone by sodium or potassium borohydride is unusual, because lactones and esters are not normally reduced by these reagents, and it is likely that the peroxy group complexes with boron to deliver hydride selectively from the back-side of the molecule to give the *β*-configured hemiacetal hydroxy group. A similar oxygen-assisted reduction of an ester to an alcohol was observed in the reduction of taxol to baccatin III by sodium borohydride [134]. The hemiacetal hydroxy group can then be converted to its methyl acetal to give β-artemether or to its ethyl acetal to give β-artemotil. Conversion to α-artesunate was originally effected by reaction with succinic anhydride and pyridine, but a later study found that the use of triethylamine in isopropyl acetate in place of pyridine gave higher yields [135]. A continuous-flow conversion of artemisinin to dihydroartemisinin, β-artemether, β-artemotil, and α-artesunate has been described [136], and a flow process such as this is likely to be the most cost-efficient method for the production of these key active pharmaceutical ingredients (APIs).

Many other derivatives of artemisinin (**18**) have been prepared, some of which have very promising activities, but none of them has yet moved into clinical use. An English-language summary of many of the derivatives prepared in early investigations in China has appeared [137], as have several other reviews on the synthesis of artemisinin derivatives [94, 132, 138]. Several promising dimeric derivatives have been developed by Posner et al. Thus, the four trioxane amide dimers **28–31** all gave 30 days of survival for mice infected with *P. berghei* after a single subcutaneous dose and with no detectable parasites remaining [139], while a similar result was obtained for mice treated with a single oral dose of the sulfone dimer **32** plus mefloquine (**16**) [140]. Thioacetal **33** in combination with **16** in a single oral dose provided a mean survival time in mice infected with *P. berghei* ANKA strain of 29.8 days, compared with 16.5 days for artemether plus **16** [141].

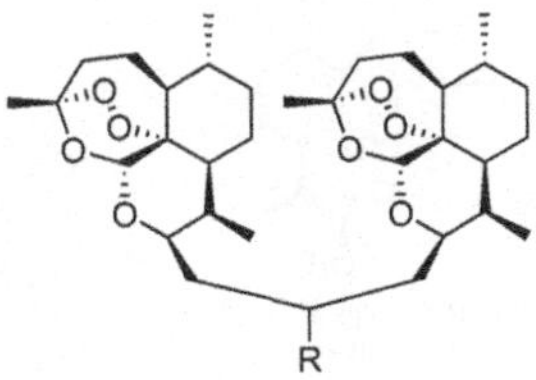

28 R = CH_2NHPh
29 R = $CONHCH_2Ph$
30 R = $CONHC(CH_3)_3$
31 R = $CONHNHCOC_5H_4N$

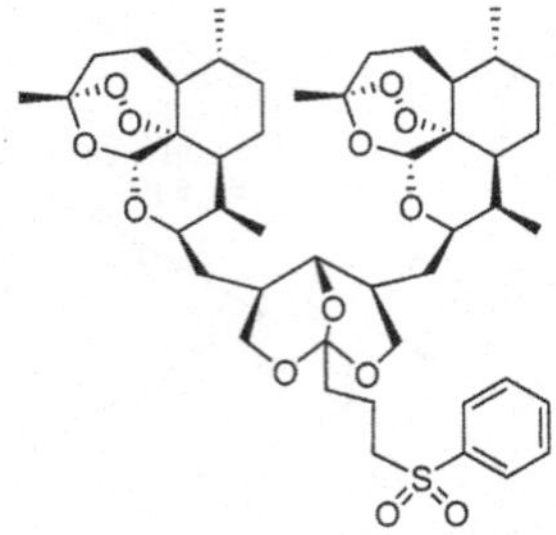

32

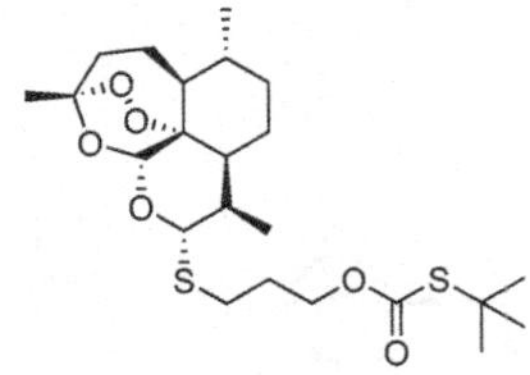

33

A recent search for new antimalarial agents by virtual screening of the MolPort commercial database of 120,000 compounds identified 28 virtual hits, of which eight were evaluated in vitro against *P. falciparum* asexual stages. The only ones found with significant activity were the artemisinin derivatives **34** and **35**, with IC_{50} values of 0.5 and 0.7 n*M* against the 3D7 strain of *P. falciparum*, slightly more potent than artesunate at 1.6 n*M* [142]. This study shows the potential of virtual screening to identify lead compounds, although the results in this case did not identify any significantly active new structural types.

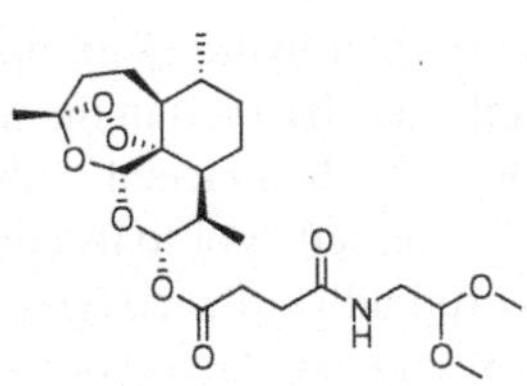

34

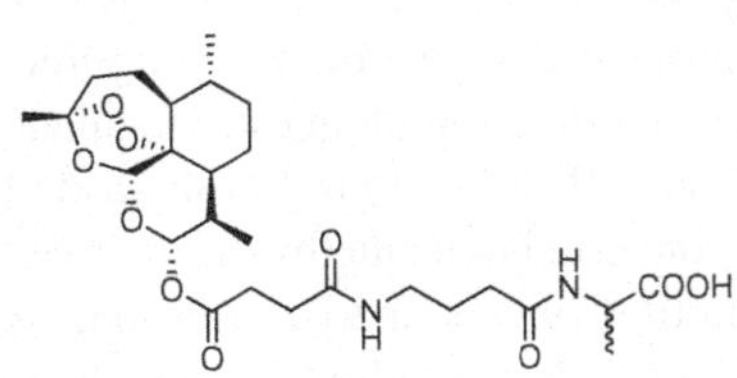

35

Several synthetic endoperoxides have also been prepared in attempts to develop synthetically accessible drugs that target malaria by a similar mechanism to artemisinin (**18**), and two such derivatives are of significant interest. The trioxane OZ277, or arterolane (**36**), was developed by a multi-institutional collaborative team with the support of the Medicines for Malaria Venture (MMV), and was selected for development on the basis of its potency and lower toxicity compared with other analogs [143]. Clinical trials were carried out by Ranbaxy and were initially disappointing, but the company persevered and obtained approval to market a combination of arterolane and piperaquine (**37**) in India in 2012 [144] and in seven African countries in 2014 [145]. The related trioxane OZ439, or artefenomel (**38**), was developed in 2011 [146]. It completely cured *P. berghei*-infected mice with a single oral dose of 20 mg/kg and exhibited prophylactic activity superior to mefloquine. It has successfully completed a Phase 2 clinical trial with the conclusion "Artefenomel is a new synthetic antimalarial peroxide with a good safety profile that clears parasitaemia rapidly in both *P falciparum* and *P. vivax* malaria. Its long half-life suggests a possible use in a single-dose treatment in combination with other drugs" [147]. Two Phase 2 trials of a combination of artefenomel with ferroquine (**39**) were completed in 2020 [148], but it is not clear whether it will be approved for clinical use, at least in part because of solubility problems [149].

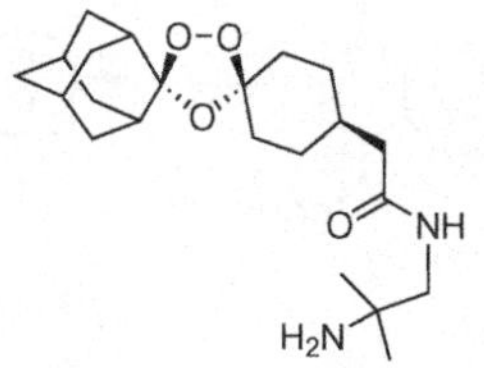

36 (OZ277, arterolane)

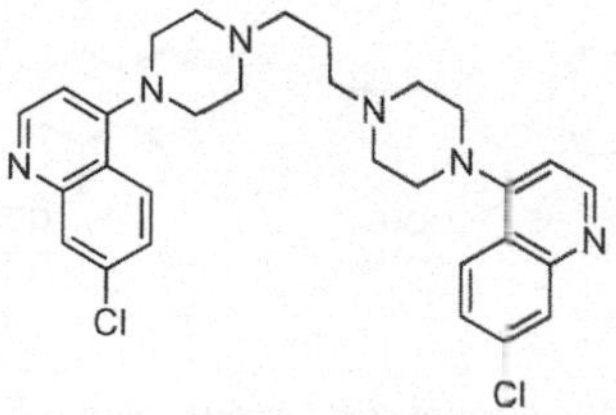

37 (piperaquine)

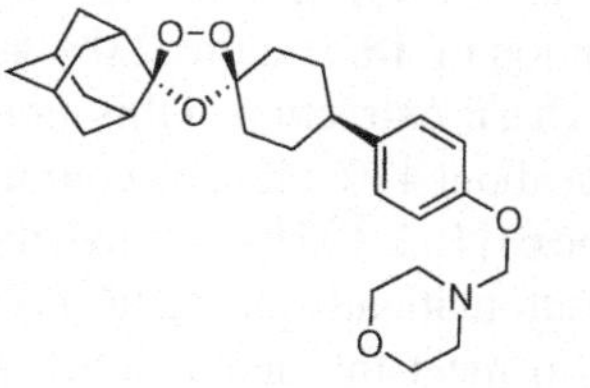

38 (OZ439, artefenomel)

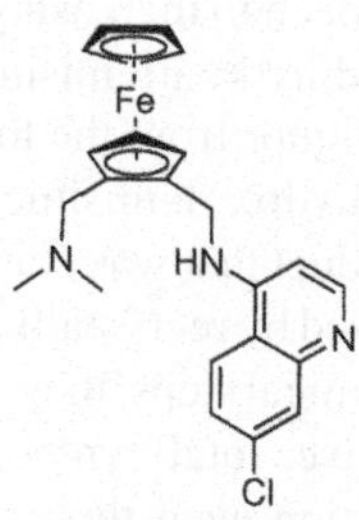

39 (ferroquine)

3.3.4 The Artemisinin Supply Problem

The three major options in meeting the need for the large-scale supply of artemisinin (**18**) and its analogs for clinical use are cultivation of *A. annua*, total synthesis, and semisynthesis from a low-cost intermediate. The same three options applied to the supply of paclitaxel (Taxol™) when it became an important anticancer drug [150], and there are significant parallels between the paclitaxel and artemisinin situations.

Current supplies of artemisinin (**18**) are provided largely by cultivation of *A. annua,* primarily in China, Vietnam, and East Africa, but the cultivation, extraction, and purification are costly, and so this life-saving drug is expensive. Due to variations in supply and demand, the cost of **18** varied from $1100 to $200 per kilo between 2004 and 2015 [116]. The Gates Foundation set a target price of $100 per kilo or less for the production of 50 metric tons of artemisinin per year in its latest request for proposals [116], and this price cannot be met using current plant cultivation methods. Various approaches to increasing artemisinin (**18**) production in *A. annua* and also in *Nicotiana tabacum* have been summarized in a review by Farhi et al. [151], in which they emphasize the need to apply state-of-the-art synthetic biology approaches to ensure successful bioproduction of the drug.

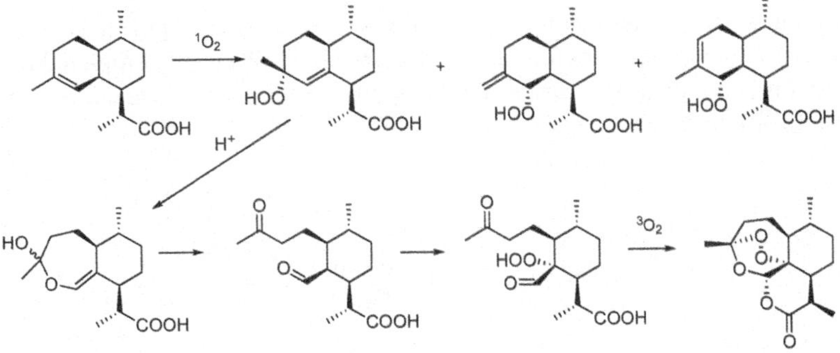

18 (artemisinin)

Scheme 6 Flow chemistry conversion of dihydroartemisinic acid to artemisinin (**18**)

Another approach to increasing the supply of artemisinin (**18**) from *A. annua* is to make use of the dihydroartemisinic acid (**23**) that is a side product from isolation of **18**. The mother liquor from the final crystallization of **18** contains 2.0% artemisinic acid and 8.2% dihydroartemisinic acid (**23**), and basic extraction of this mother liquor gave a crude product that was converted to **18** in about 41% yield by continuous flow chemistry followed by recrystallization (Scheme 6) [152, 153]. Although the reaction requires five chemical steps, they all take place automatically during the flow process.

As noted above, total synthesis is unable to meet the demands of large-scale low-cost production even though it is innovative and can give reasonable overall yields. In the case of paclitaxel, the supply problem was initially met by aggressive *Taxus brevifolia* bark collection, but the medium-term need was met by semisynthesis from the more readily available 10-deacetylbaccatin III (Scheme 7) [154]. Current production of paclitaxel is by a combination of isolation from various *Taxus* species and by plant tissue culture [155], some of which is carried out by Phyton Biotek in Germany.

A plant tissue culture approach to artemisinin (**18**) production has not yet proved commercially feasible, but several investigators have studied the effects of growth conditions on **18** production in *A. annua* tissue cultures. One of the first was a study by Woerdenbag et al., who examined the effects of additives such as gibberellic acid on artemisinin production in shoot cultures of *A. annua* [156]. Baldi and Dixit studied cell cultures of *A. annua*, and were able to produce yields of 110 mg/dm^3 by

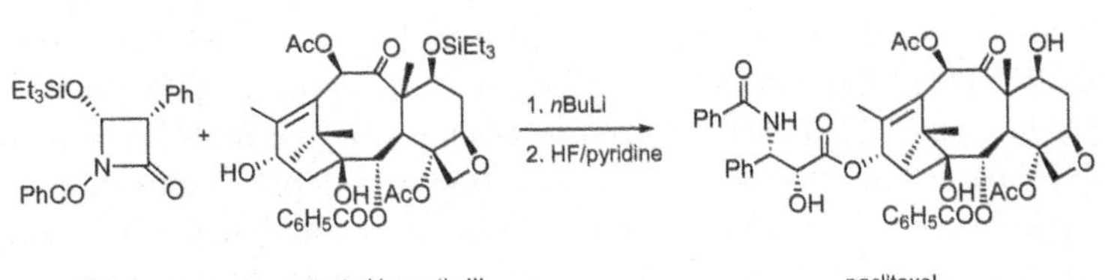

β-lactam protected baccatin III paclitaxel

Scheme 7 Semisynthesis of taxol from baccatin III

using mevalonic acid lactone as the precursor and methyl jasmonate as an elicitor [157]. Impressive as this work is, the use of expensive mevalonic acid lactone as the precursor limits its commercial viability. Phyton Biotek is working on developing a cell culture approach with the support of the Gates Foundation [158], but the outcome of this approach is not yet known. A review of artemisinin production in plants concludes with the hopeful statement "Perhaps the continuous production of artemisinin and other valuable plant metabolites in suspended bioreactor cultures with in situ extraction to avoid cell toxicity is not too far in the future. This will avoid the regulatory restrictions on in field GMO plants, and allow for stable continuous production of drugs" [109].

Hairy root cultures have also been studied as a source of artemisinin (**18**). Hairy roots can be induced by the soil bacterium *Rhizobium rhizogenes*, and are easy to culture, but suffer from the difficulty of scale-up to commercial production because of their delicacy. A study by Liu et al. of light irradiation on artemisinin production in hairy root cultures found that production increased with 3000 lx for 16 h to yield 244 mg/dm^3 after 30 days [159]. Hairy root propagation conditions have also been studied by Srivastava and Patra. Their initial study gave yields of only 2 mg/dm^3 after 25 days of batch cultivation [160], but a later paper described an improved process involving cultivation of hairy roots in a nutrient mist bioreactor that gave yields of artemisinin of 25.8 mg/dm^3 over 20 days [161]. A review of approaches to the production of **18** by hairy root cultures has appeared [162].

Fortunately, synthetic biology approaches to the semisynthesis of artemisinin (**18**) have begun to yield fruit, and a first-generation approach based on work by Keasling has made a modest impact on the supply problem. This synthetic biology approach was built on earlier work on the production of amorphadiene in *E. coli* [163], which was later followed by work on the production of amorphadiene in yeast and chemical conversion of the diene to artemisinic acid to give a final yield of 1.6 g/dm^3 of artemisinic acid [164]. An improved process overexpressed all the genes for the production of artemisinic acid in yeast (Scheme 2), as well as repressing the *ERG9* gene for squalene synthase, and increased the yield of the acid more than an order of magnitude, to 25 g/dm^3 [165–167]. This process was selected for commercial production by Sanofi at its production facility in Garessio, Italy after improvements to the process chemistry for the conversion of artemisinic acid to artemisinin (**18**). These improvements increased the overall yield from 19 to 55% [168], and gave overall production costs of about $400/kg. Semisynthetic artemisinin (**18**) produced by this route accounted for about 25 and 15 metric tons of **18** in 2013 and 2014 respectively, but none in 2015, because the agricultural cost was less than $400/kg from 2013 to 2015. Sanofi then sold their Garessio plant to Huvepharma, which produced about 20 metric tons of **18** in 2017 using enhanced technology [116].

An improved but not yet commercial production method takes the counterintuitive step of using synthetic biology to produce amorphadiene (**19**) in yeast, and then using chemistry to convert the diene to artemisinin (**18**) (Scheme 8). Although additional chemical steps are needed to convert amorphadiene to artemisinic acid en route to **18**, **19** is produced in up to tenfold larger amounts than artemisinic acid in yeast [164],

Scheme 8 Conversion of amorphadiene (**19**) to artemisinic acid

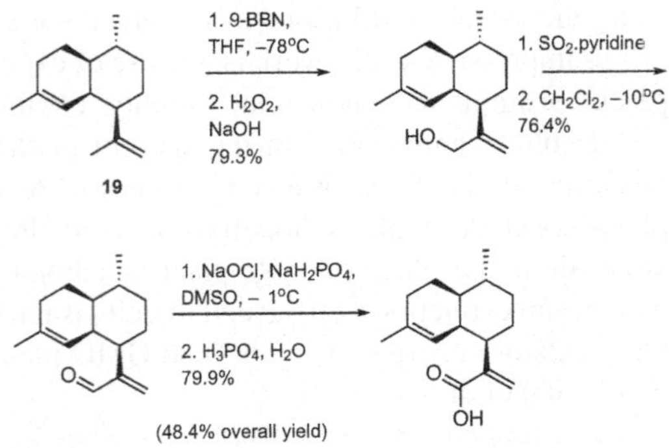

so the additional chemical steps are more than worthwhile from an overall cost point of view.

The initial synthesis of artemisinic acid from amorphadiene (**19**) (Scheme 8) required the use of expensive 9-BBN as a key reagent [164], but an improved synthesis has been developed that uses only inexpensive reagents to convert **19** to dihydroartemisinic acid (**23**) stereoselectively [169] (Scheme 9) and thence to artemisinin (**18**) by chemistry developed by Chen et al. [170] in an overall yield from amorphadiene (**19**) to artemisinin (**18**) of 40%. The key to this synthesis is the "protection" of the internal double bond of **19** by epoxidation, followed by a three-step oxidation of the side chain and stereoselective reduction of the external double bond with diimine. If this pathway can be scaled to commercial production it appears to offer a lower cost process for artemisinin generation [169], and is superior to earlier work by Haynes and Vonwiller, who were able to convert artemisinic acid to dihydroartemisinin in 46% yield [171].

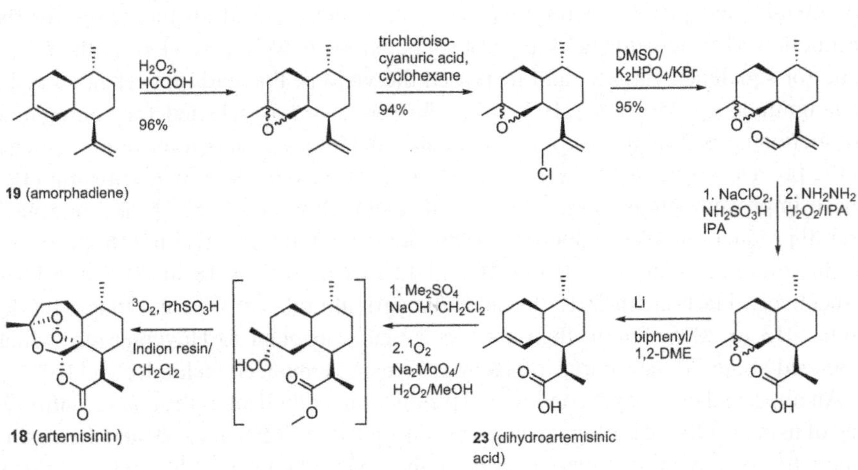

Scheme 9 Conversion of amorphadiene to artemisinin (**18**)

A recent synthesis by Wirth and Allemann used the enzyme amorphadiene synthase to convert farnesyl diphosphate to amorphadiene, and then either chemistry or an engineered yeast to convert this to dihydroartemisinic acid (**23**) [172]. The derived methyl dihydroartemisinate was then converted to artemisinin (**18**), but in very low yield, and the cost of the farnesyl diphosphate starting material is also an issue, so this route will need significant improvement to become commercially viable.

Several other teams are working on the bioproduction of artemisinin (**18**). These include Manus Bio, which recently was awarded a next stage of funding from the Gates Foundation to continue its development of a microbial route to artemisinic acid [173], and reviews on approaches to the commercial production of semi-synthetic **18** have appeared [174, 175]. The latter review concludes "Industrial production of semi-synthetic artemisinin began in 2013, but technological advances described above in both biology and chemistry have opened opportunities to improve the process and decrease the cost of semi-synthetic artemisinin production. Developing these technologies could further safeguard the supply of artemisinin for those who need it most in the developing world" [175].

3.3.5 Mechanism of Action

The mechanism of action of artemisinin (**18**) and related compounds is not yet completely settled, in spite of numerous studies over the last 30 years. What is abundantly clear is that the ozonide (trioxane) portion of the structure is required for potent activity, since deoxyartemisinin is 250-fold less potent than **18** against the D-6 clone of *P. falciparum* [176], and since synthetic ozonides such as arterolane (**36**) and artefenomel (**38**) have strong antimalarial activity. In addition, the mechanism must involve free radicals at some level, since free-radical scavengers such as catalase and dithiothreitol inhibit artemisinin's antimalarial activity [177]. Finally, the mechanism must also involve iron in some way, since iron chelators such as pyridoxal benzoyl hydrazine reduce the in vitro antiplasmodial activity of artemisinin, and the activity requires hemoglobin uptake and digestion [178, 179]. Beyond agreement on these issues, there are several major schools of thought on the mechanism of artemisinin (**18**). Proposed theories up to 2004 are described in reviews by Meshnick [180] and O'Neill and Posner [181], and later reviews by Haynes et al. [182], Liu [102], and Kumari et al. [132] provide alternate mechanisms.

The Carbon Radical Hypothesis

The carbon radical model (Scheme 10) focuses on the conversion of hematin to heme in the parasite food vacuole lipid bilayer as the source of ferrous iron that then reacts with the endoperoxide bridge of artemisinin (**18**) to generate carbon radicals such as **40** and **41**. These radicals are then proposed to react intramolecularly to generate stable products, or with target biomacromolecules such as heme or nearby proteins

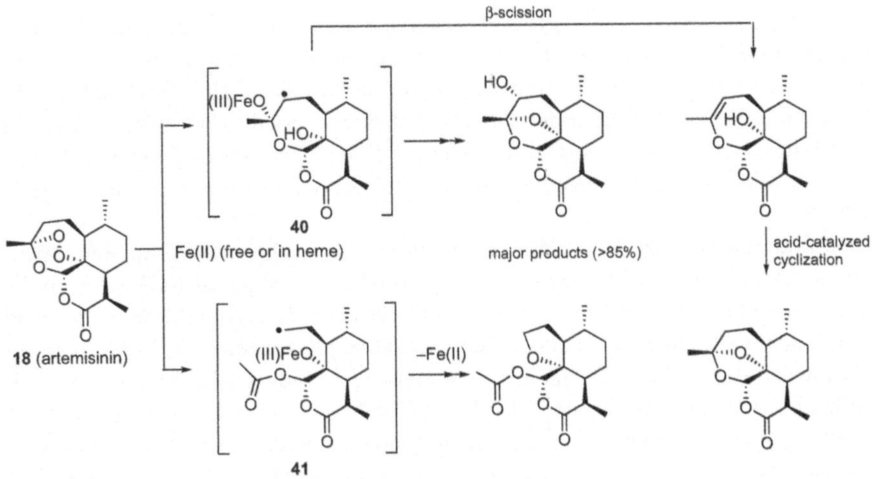

Scheme 10 Homolytic cleavage of the endoperoxide bridge of artemisinin (**18**) to generate carbon-based radicals. Adapted from O'Neill and Posner [181]

[183]. This model is supported by a study showing that **18** enhances heme-catalyzed oxidation of lipid membranes [184].

This hypothesis has been challenged by Wu, who argues that the intramolecular formation of the major products would be faster than the intermolecular reaction with biomacromolecules, and that free radicals would be eliminated rapidly by glutathione and the antioxidative stress enzyme system [185]. It has also been challenged by Haynes on the basis of his finding that the synthetic compound **42**, unlike artemisinin (**18**), is inert to ferrous heme but is completely decomposed by free Fe(II), and nevertheless has superior in vitro activity against *P. falciparum* [186]. In addition, a series of aminoartemisinins of general structure **43** have potent antimalarial activities but are only poorly converted into C-radicals [187]. Haynes concludes his evaluation of the evidence for the carbon-based radical hypothesis with the words "Overall, it must be concluded that the C-radical hypothesis as the basis for antimalarial activities of artemisinins and synthetic peroxides is untenable" [182].

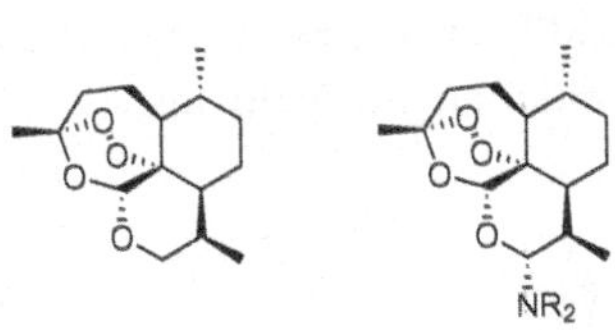

42 **43**

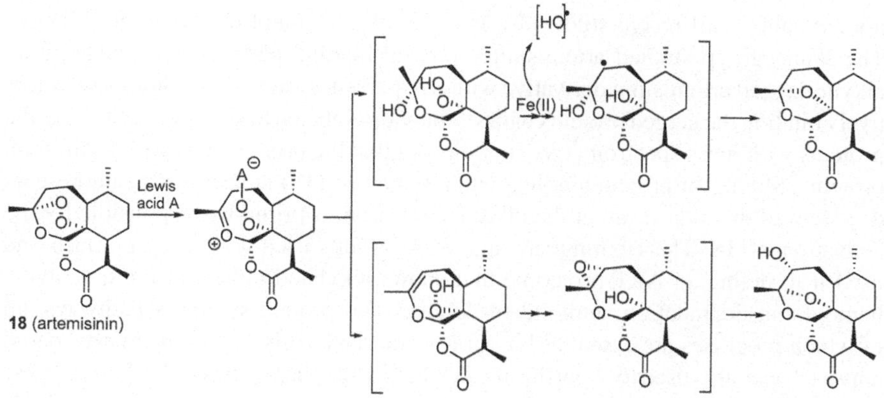

Scheme 11 Heterolytic cleavage of the endoperoxide bridge of artemisinin (**18**). Adapted from O'Neill and Posner [181]

The Oxygen Radical Hypothesis

An alternative proposed heterolytic cleavage could generate a ring-opened compound by protonation or more likely Lewis acid-catalyzed opening of the peroxide, which could then add water and generate hydroxyl radicals by Fenton chemistry in the presence of iron (Scheme 11). The hydroxyl radical in particular is known to degrade biomacromolecules, and oxidative stress is a major factor in the high mortality rate of various diseases, including malaria [188].

The SERCA Hypothesis

The malaria parasite sarco-endoplasmic reticulum Ca^{2+} (SERCA) ATPase PfATP6 is a key enzyme for parasite growth and is the only SERCA-type ATPase in the parasite's genome. Studies with PfATP6 expressed in *Xenopus laevis* oocytes showed that artemisinin (**18**) inhibited PfATP6 with similar potency to the sesquiterpene lactone thapsigargin [189]. However, a claim that thapsigargin antagonized the parasiticidal activity of artemisinin was later disproved [190], and a docking study that included **18** and its derivatives into the thapsigargin binding cleft showed no correlation between the binding affinity of these drugs to PfATP6 and their antimalarial activities [191]. It thus appears that although PfATP6 may be a target of artemisinin (**18**), it cannot be the only target and probably is not a major target.

The General Protein Binding Hypothesis

The binding of artemisinin (**18**) to parasite proteins was first observed by Asawamahasakda et al. [192], but information on the identities of the bound proteins was not available until recent studies by Wang et al. [193] and by Ismail et al. [194]. The Wang group studied artemisinin's protein binding partners by incubating an alkyne-tagged artemisinin derivative with live parasites, attaching a biotin tag, selectively binding the tagged proteins with streptavidin-labeled beads, and identifying the proteins with mass spectrometry. This showed that the probe bound to 124 different proteins. Studies of protein binding with artemisinin (**18**) labeled with a fluorescent dye showed that heme, rather than free ferrous iron, is primarily responsible for the activation of **18**. The Hemingway and Ward groups used a similar approach and used it to define an ART proteome that shows alkylated targets in the glycolytic, hemoglobin degradation, antioxidant defense, and protein synthesis pathways, all of which processes are essential for parasite survival. This "click-chemistry" based approach was also used to show that the synthetic trioxolane drugs OZ277 and OZ439 alkylate a similar set of proteins [195].

The Oxidative Stress Hypothesis

An alternative mechanism proposed by Haynes and Monti is that enhancement of oxidative stress mediated by peroxidic antimalarials such as artemisinin (**18**) leads to loss of redox homeostasis. This proposal is based on studies that show that both artemisinin and related compounds, including synthetic peroxide analogs, can overwhelm redox homeostasis in the malaria parasite [196]. In a later review, Haynes discusses both the radical hypotheses described above in Schemes 9 and 10 and the heme hypothesis, in which the peroxides are not the active agents, but exert their parasiticidal effects through allowing the build-up of free heme-Fe^{3+} [182]. He concludes that the C-radical hypothesis is not tenable as the basis for the antimalarial activities of artemisinins and synthetic peroxides, and that the stage specificities of artemisinins compared with those of 4-aminoquinolines and arylmethanols suggest that heme is not the target, although hemoglobin digestion by the parasite is required for artemisinin-induced increase in oxidative stress [179].

Other Activities

In addition to its activity as an antimalarial agent, artemisinin (**18**) and its analogs have been found to have important other activities, including significant anticancer activity leading to clinical trials [197–200] and antiviral activity [201]. HSP90 has been identified as a direct target of artemisinin for its anti-inflammatory activity via quantitative chemical proteomics [202]. This anti-inflammatory activity has made it a potential candidate for treatment of COVID-19, but concerns have been raised that this would overwhelm the supply chain and deprive malaria victims of a crucial medicine [203].

Artemisinin Conjugates

In addition to the artemisinin derivatives discussed earlier, various conjugates of artemisinin (**18**) with other antimalarial agents have been prepared. As one example, **18** has been conjugated with the phytohormone endoperoxide G-factor analog to give hybrids **44** and **45** [204]. These compounds had IC_{50} values against the FcB1/Columbia strain of *P. falciparum* of 2.6 and 8.4 n*M*, respectively, with selectivity indices (SIs) compared with the primary human fibroblast cell line AB943 of 16,372 and 2108. For comparison purposes, artemisinin's values were 72 n*M* and >4545.

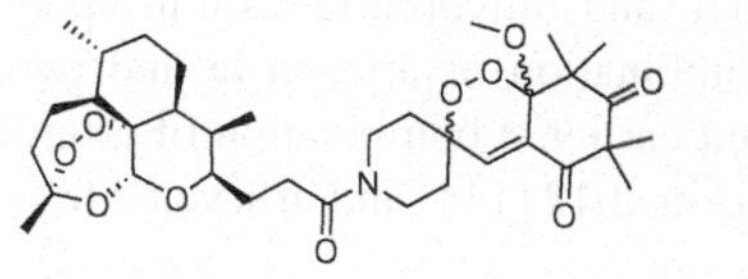

44

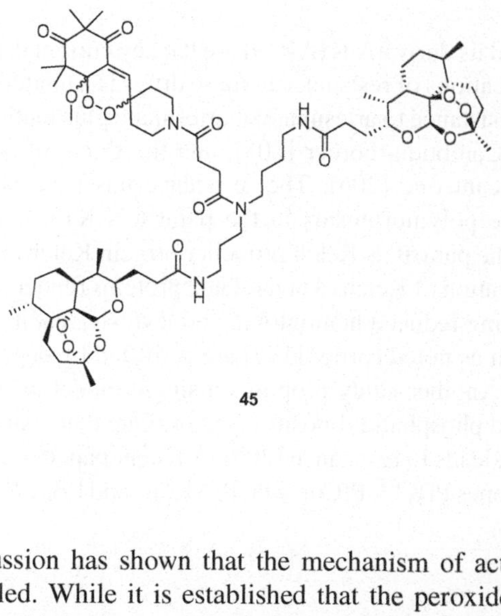

45

Conclusion

The previous discussion has shown that the mechanism of action of artemisinin (**18**) is not yet settled. While it is established that the peroxide bond is essential for potent activity and it is also established that heme is not the target, many other mechanistic aspects are yet to be determined. It does seem clear, however, that **18** is likely to exert its antimalarial effect by more than one mechanism, and that some combination of general protein binding and oxidative stress will emerge as the most probable mechanism.

3.3.6 Clinical Use and Resistance

Clinical Use

Although artemisinin (**18**) was the initial drug of choice for treatment of malaria, it has since been replaced by combination therapies in efforts to slow the development of drug resistance. In 2015, the World Health Organization recommended five artemisinin combination therapies (ACTs) for the treatment of uncomplicated *P. falciparum* malaria: artesunate and amodiaquine (Coarsucam or ASAQ), artesunate and mefloquine (Artequin or ASMQ), artesunate and sulfadoxine/pyrimethamine (Ariplus or Amalar plus), artemether and lumefantrine (Coartem Riamet, Faverid, Amatem, Lonart or AL), and dihydroartemisinin-piperaquine (Duo-Cotecxin, or Artekin) [191]. The combination of artesunate and pyronaridine (Pyramax) was added in 2017. As noted earlier, a combination of arterolane and piperaquine was approved for use in India in 2012 [144] and in seven African countries in 2014 [145].

Resistance to Artemisinin and Analogs

Artemisinin (**18**) and its derivatives (ARTs) are the key current drugs for treatment of malaria, so any indications of resistance to these drugs is a matter of serious concern. The first report of resistance to artesunate or artesunate plus mefloquine came from a study on the Thai–Cambodia border [205], and the threat of artemisinin-resistant malaria is a significant one [206]. The resistance has been shown to be caused by single-nucleotide polymorphisms in the parasite's K13 gene [207] leading to point mutations in the parasite's Kelch propeller protein Kelch13. It has been shown recently that deactivation of Kelch13 and related proteins rendered parasites resistant to ARTs by displaying reduced hemoglobin endocytosis, leading to reduced activation of ARTs, which as noted earlier [179] are activated by degradation products of hemoglobin [208]. Another study proposes a single model of resistance involving elevation of the lipid phosphatidylinositol-3-phosphate that results in vesicle expansion, which, in turn, leads to resistance [209]. A recent paper confirms that polymorphisms in the four genes PfK13, PfCoronin, PfAP2 μ, and PfUBP1 confer artemisinin resistance [210].

3.4 Other Clinically Used Natural Products

3.4.1 Tetracycline and Doxycycline

The microbial antibiotics tetracycline (**46**) and its semisynthetic analog doxycycline (**47**) have been used both for malaria prophylaxis as well as treatment, and doxycycline is used in Southeast Asia in combination with quinine for malaria treatment [211]. Their antibacterial activity is due to inhibition of bacterial protein synthesis

[212], but their mechanism of action against *Plasmodium* spp. appears to be due to their ability to block expression of the apicoplast genome, leading to a slow but potent antimalarial effect [213]. A 2011 report examines the evidence for the use of doxycycline (**47**) for malaria and summarizes the available literature on its safety and tolerability [214].

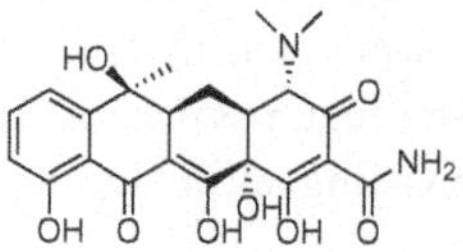

46 (tetracycline)

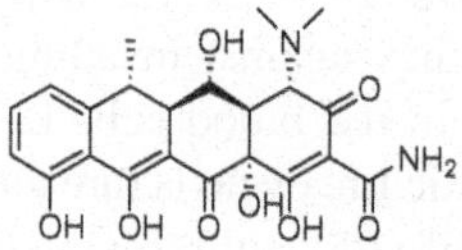

47 (doxycycline)

3.4.2 Clindamycin

Lincomycin (**48**), an antibiotic isolated from *Streptomyces lincolnensis,* and its chlorinated semisynthetic derivative clindamycin (**49**), are both used clinically, although **49** is the more important of the two [215]. As is the case with tetracycline (**46**) and doxycycline (**47**), both drugs act as protein synthesis inhibitors, and their antimalarial activity is due to their ability to shut down protein synthesis in the apicoplast [216]. A 2002 review of the clinical use of clindamycin (**49**) indicated that although it is not an ideal drug because of its short half-life and the concomitant need for twice daily dosing, its combination with quinine is "an extremely useful antimalarial combination" for children and pregnant women [217]. A recent review by Saito et al. stated "Quinine plus clindamycin, therefore, can be an option when artemisinin cannot be used (e.g. allergy), but special care (e.g. full supervision, and close clinical and parasitological monitoring with hospitalization) should be considered" [218].

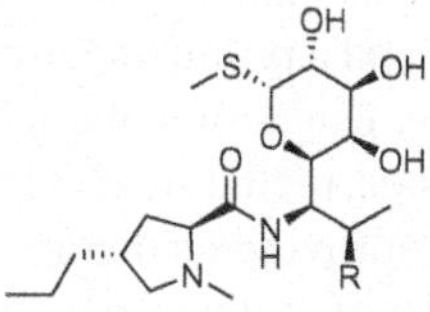

48 R = H (lincomycin)
49 R = Cl (clindamycin)

3.4.3 Azithromycin

Azithromycin (**50**) is a natural product derivative that was prepared in 1980 by chemical modification of the antibiotic erythromycin [219]. Like clindamycin (**49**), it is an inhibitor of protein synthesis in the apicoplast, and it has been shown to be an effective antimalarial agent in combination with quinine [220]. However, the Cochrane database concludes "azithromycin is a weak antimalarial with some appealing characteristics. Unless it overcomes some of the highlighted challenges and finds a specific niche that is complementary to the current scala of more efficacious antimalarial combinations, its future as an antimalarial does not look promising" [221]. Recently, members of a series of 84 azithromycin analogs were identified with nanomolar quick-killing potency against invading merozoites and the ring stages of parasite development within red blood cells [222]. The feature of targeting early stages of the intraerythrocytic life cycle is rare among existing antimalarials, so this discovery has brought new life for azithromycin (**50**).

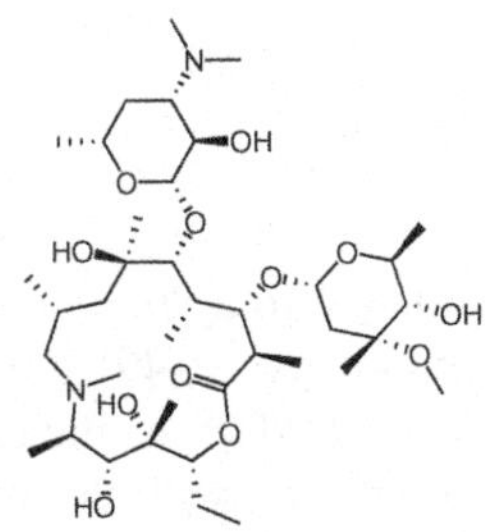

50 (azithromycin)

3.4.4 Fosmidomycin

Fosmidomycin (**51**) is a natural antibacterial agent originally isolated from *Streptomyces lavendulae* in the late 1970s [223]. Its antimalarial activity is due to inhibition of 2*C*-methyl-D-erythritol synthase, which is the first committed step of the non-mevalonate isoprenoid biosynthesis pathway, known as the methylerythritol phosphate pathway in the apicoplast, and a promising drug target since it is not present in mammals [224]. It has been investigated for malaria treatment both as a monotherapy and in combination with other drugs, including clindamycin, and several analogs have been prepared with the goal of improving its therapeutic profile [225]. A 2015 meta-analysis of clinical data concluded that it was not yet ready for wide use, but that its combination with clindamycin was still worth investigating [226].

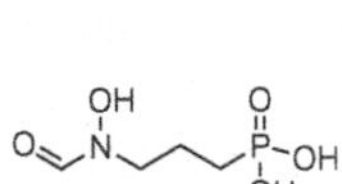

51 (fosmidomycin)

3.5 *Natural Products with Potent Antiplasmodial Activity*

3.5.1 Introduction

The practical constraints on the discovery of new antimalarial natural products are severe, since cost is a significant factor for antimalarial drugs, and many natural products occur in relatively low amounts in their host organism. As an example, the so-called "taxol supply crisis" [150] was only solved successfully because of the high value of taxol as an anticancer drug; such an expensive effort would never have been considered for an antimalarial drug.

The natural products discussed in this section have been selected on the basis of their cell-based activity as antiplasmodial agents, in displaying submicromolar antiplasmodial activity against both chloroquine-sensitive and chloroquine-resistant strains of *P. falciparum*, and at least a 100-fold differential between activity against a chloroquine-resistant strain and mammalian cytotoxicity, although a few examples of interesting compounds that do not meet these criteria are also included. These criteria eliminate the many compounds with weak or moderate activity that are reported regularly in the scientific literature, and it is hoped make the present contribution a useful source of curated information on the current status and promise of new potential antimalarial agents from Nature.

For a hit compound with good antiplasmodial activity and selectivity to become a realistic lead compound, it must also be both orally bioavailable and effective in an animal model. These qualities are normally evaluated by measuring a compound's effectiveness in reducing parasitemia in mice infected with *P. berghei*. The term "antimalarial activity" is used in this chapter for the relatively few compounds that meet these criteria, and the term "antiplasmodial activity" is used for the remaining hit compounds. Compounds with reported in vivo activity are included even if their antiplasmodial activity does not meet the above criteria.

A variety of different strains of *P. falciparum* have been used as test strains, some of which are chloroquine-sensitive and others are chloroquine-resistant. These strains are summarized in Table 1.

The natural products are categorized by their structural class, regardless of their reported biological origin as microbial, marine, or plant derived, since most marine natural products [243] and even some plant natural products have been shown to have microbial origins. However, the majority of natural products with confirmed antiplasmodial activity are plant products, both because of the long history of the use of plant products as herbal remedies for malaria and because of the relatively easier access to plants as opposed to microbial and marine sources.

Several comprehensive reviews of antimalarial natural products have appeared in the past few years, and these are referenced here for readers desiring information on compounds with weaker activities than those discussed herein. An excellent comprehensive review by Saraiva and Dimopoulos on antiplasmodial natural products from bacterial sources was published in 2020 [244], and another excellent and comprehensive review by Tajuddeen and van Heerden appeared in 2019 [245] covering

Table 1 Standard *Plasmodium falciparum* strains used for in vitro antimalarial efficacy studies[a]

Strain	Origin	Resistant to	References
Dd2	Indochina	Chloroquine, pyrimethamine, mefloquine	[227]
W2	Indochina	Chloroquine	[228]
HB-1–2–3	Honduras		[229]
3D7	Netherlands		[230]
NF54	Netherlands		[231]
D6	Sierra Leone		[232]
D10	Papua New Guinea		[233]
Malayan Camp	Malaysia	Proguanil, cycloguanil	[234]
FCR-3/FMG	Gambia	Chloroquine	[235]
7G8	Brazil	Pyrimethamine	[236]
K1	Thailand	Multidrug	[237]
GB4	Ghana	Chloroquine	[238]
Sta. Lucia	Santa Lucia		[239]
V1/S	Vietnam	Chloroquine, quinine	[240]
FVO	Vietnam	Chloroquine	[241]
FC27/PNG	Papua New Guinea		[242]

[a]The authors thank Dr. Emilio F. Merino for assistance in preparing this table

the literature on antiplasmodial natural products reported between 2010 and 2017. Other reviews of note include a 2002 review by Schwikkard and van Heerden [246] embracing the literature from 1990 to 2000, a 2010 review by Wright covering the literature from 2000 to early 2010 [247], a second review by Wright published in 2015 [248], and a 2018 review by Pan et al. [249] dealing with the literature from 2001 to 2017. These reviews have been very helpful in the preparation of the present contribution. Other reviews include those by Bekono et al. on antimalarial compounds derived from African medicinal plants [250], by Silva et al. on antimalarial terpenes [251], by Kaur et al. [252], and by Nair and van Staden on antiplasmodial lycorine alkaloids [253] and crinane alkaloids [254].

This contribution focuses on selected natural product antiplasmodial agents reported up to 2017, with a somewhat broader coverage of compounds reported from 2018 to early 2021. Interested readers are referred to the reviews cited above for more extensive coverage of antimalarial plant products reported before 2018.

3.5.2 Alkaloids

Simple Alkaloids

The novel aromatic alkaloid cassiarin A (**52**) was isolated from the leaves of *Cassia siamea*, and had potent antiplasmodial activity with an IC_{50} value of 23 n*M* against an unspecified strain of *P. falciparum* and a good SI of 1380 against the P388 mammalian leukemia cell line [255]. It also has in vivo activity against *P. berghei* in mice [256]. Cassarin A (**52**) has been synthesized in 52% yield from methyl 2,4-dihydroxy-6-iodobenzoate [257], in 39% yield from *N*,2,4-trimethoxybenzamide [258], and in 53% yield from 1-(bromomethyl)-3,5-bis(methoxymethoxy)benzene [259], so larger amounts are potentially available. Various analogs of cassiarin A (**52**) have been synthesized [260, 261], and although none of them appears to have improved antimalarial activity compared with the natural product, some of the derivatives have cytotoxicity, DNA binding and topoisomerase-II inhibition activities [260], and the natural product and a methoxy derivative had significant vasorelaxant activity [261]. Cassiarin A (**52**) is thus a potentially interesting lead compound for development.

Quinolone **53** was isolated in low yield from *Crinum firmifolium*, and had a moderate antiplasmodial activity of 0.58 μ*M* and a SI of 58 against the A2780 ovarian cancer cell line. Both the activity and selectivity were improved in a synthetic analog **54**, with an IC_{50} of 0.33 μ*M* and a SI of 89 [262].

Acridone **55** was synthesized along with over 90 related compounds based on acridone alkaloid natural product models. It is included here because of its outstanding antiplasmodial activity and selectivity, with IC_{50} values of 0.028 n*M* against the D6 *P. falciparum* strain, 0.045 n*M* against the Dd2 strain, and 0.042 n*M* against the 7G8 strain. It also has an outstanding SI of > 1,000,000 for *P. falciparum* over HepG2 hepatic carcinoma cells. Both **55** and various analogs were evaluated against *P. yoelii* in mice, and **55** was curative at an oral dose of 10 mg/kg/day over four days. It was selected as a lead compound for development based on these results, as well as its prevention of in vitro *P. berghei* sporozoite-induced development in human hepatocytes at a low nanomolar concentration, its full in vivo protection and cure via oral administration against liver-stage *P. berghei* sporozoite-induced infection in mice, its low cardiotoxicity, and other factors [263].

Anibamine (**56**) and anibamine B (**57**) were isolated from *Aniba citrifolia* and found to have moderate antiplasmodial activities against the Dd2 strain of 0.17 and 0.24 μ*M*, respectively, but with only poor selectivity against the A2780 ovarian cancer cell line. It is noteworthy that both antiplasmodial and cytotoxic activities were dependent on the charge on the quaternary nitrogen, as the uncharged synthetic analog **58** was essentially inactive. Interestingly, the charged compounds have substantial activity against ring stages [264].

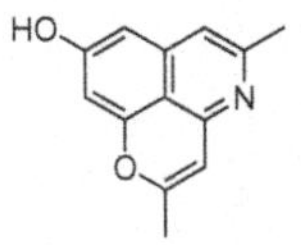

52 (cassiarin A9)
Pf 23.4 n*M*
P388 32.4 μ*M*
SI 1380

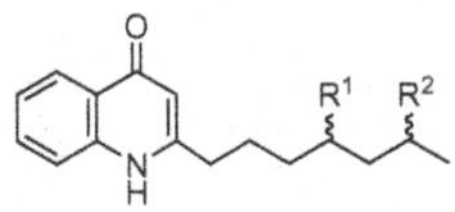

53 R^1 = H, R^2 = Me
Dd2 0.58 μ*M*, 3D7 0.50 μ*M*
SI (A2780) 58
54 R^1 = Me, R^2 = H
Dd2 0.33 μ*M*, 3D7 0.38 μ*M*
SI (A2780) 89

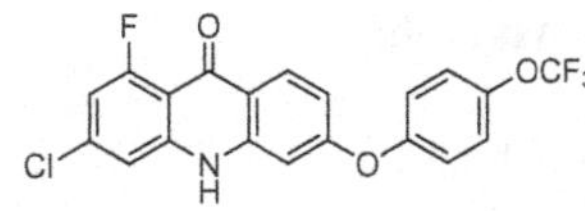

55
D6 0.028 n*M*
Dd2 0.045 n*M*
7G8 0.042 n*M*
HepG2 >35,700 n*M*
SI >1,000,000

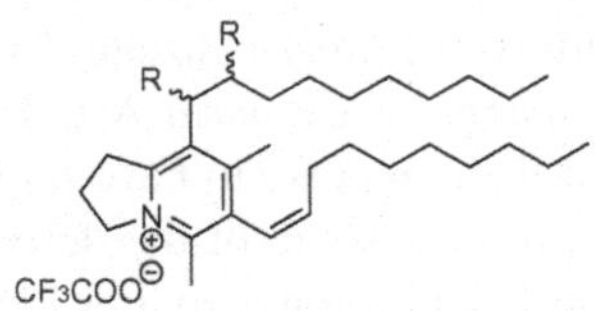

56 R = bond (anibamine)
Dd2 0.17 A2780 0.223 μ*M*
57 R = H (anibamine) B
Dd2 0.24 A2780 0.24 μ*M*

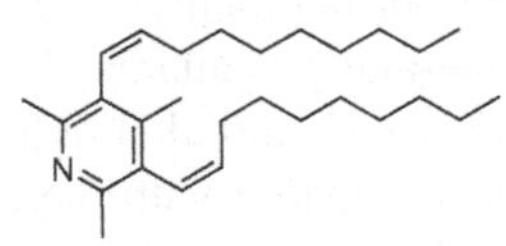

58
Dd2 20 μ*M*

Naphthylisoquinoline Alkaloids

Bringmann and his group have published extensively on the identification of new antimalarial naphthylisoquinoline alkaloids, of which many have potent and selective activity against *P. falciparum.* Two of the most promising compounds are dioncophylline C (**59**) and dioncopeltine A (**60**) from *Triphyophylum peltatum.* The former had an IC_{50} of 38 n*M* against the NF54 strain of *P. falciparum* and had 550-fold selectivity for *P. falciparum* over HepG2 cells. It was shown to cure mice infected with *P. berghei* by oral administration with minimal toxicity; its ED_{50} value was 10.7 mg/kg/day. Dioncopeltine A (**60**) was even more potent against NF54 cells with an IC_{50} of 8 n*M* and an SI of 4,890. It was able to suppress parasitemia almost totally [265]. In a later study, it was demonstrated that dioncophylline C (**59**) forms a complex with ferriprotoporphyrin IX heme in solution [266], a finding that could open this class of compounds to further development.

The atropo-diastereomeric alkaloids ancistectorine A2 (**61**) and 5-*epi*-ancistectorine A2 (**62**) also have potent antiplasmodial activity against the chloroquine-resistant K1 strain of *P. falciparum*, with IC_{50} values of 70 and 30 n*M*, respectively, and SIs against L6 rat myoblast cells of 700 and 3,300 [267]. Dioncophylline F (**63**) from the Congolese liana *Ancistrocladus ileboensis* was active against the K1 and NF54 strains with IC_{50} values of 45 and 90 n*M*, respectively, and SIs against L6 cells of 300 and 160. Dioncophylline F (**63**) was also synthesized by Pd-catalyzed Suzuki–Miyaura cross-coupling of a bromoisoquinoline and a naphthylboronic acid ester [268], potentially making this compound available on a large scale. Habropetaline A (**64**) was isolated from *Triphyophyllum peltatum*, and had a strong antiplasmodial activity of 15 n*M* against the NF54 strain and an SI of 2900 for NF54 over HepG2 cells [269].

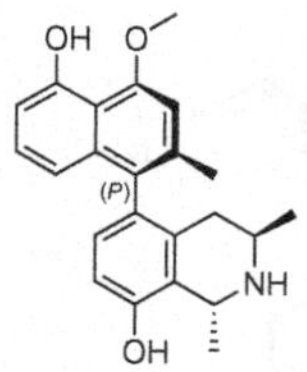

59 (dioncophylline C)
NF54 38 n*M*
SI (HepG2) 550

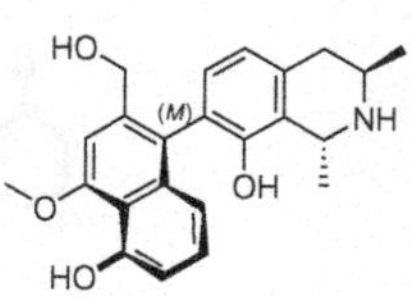

60 (dioncopeltine A)
NF54 8 n*M*
SI (HepG2) 4890

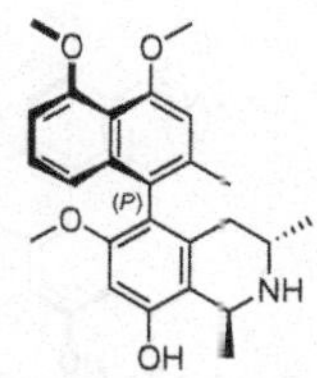

61 (ancistectorine A2)
K1 70 n*M*
L6 49 μ*M*
SI 705

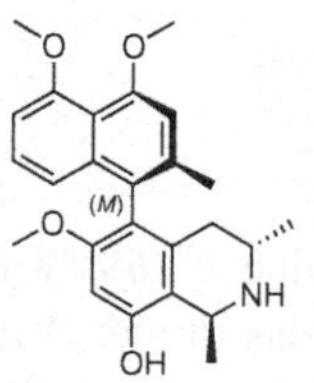

62 (5-*epi*-ancistectorine A2)
K1 30 n*M*
L6 100.1 μ*M*
SI 3340

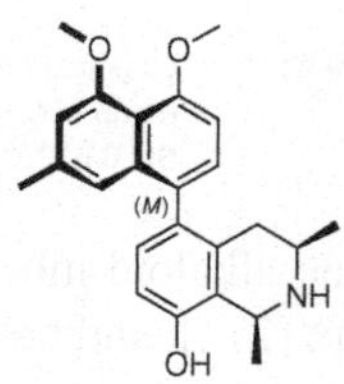

63 (dioncophylline F)
K1 45 n*M*
SI 323 (L6)
NF54 90 n*M*
SI 161 (L6)

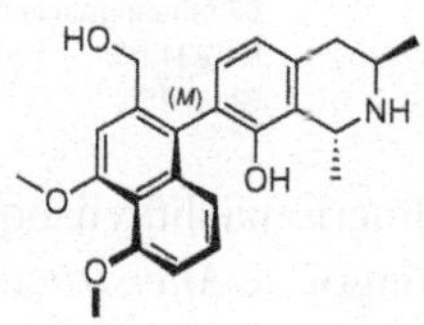

64 (habropetaline A)
NF54 15 n*M*
SI (HepG2) 2900

Shuangancistrotectorines A–D (**65**–**68**) are dimeric naphthylisoquinoline alkaloids isolated from the Chinese plant *Ancistrocladus tectorius*, and are unusual in having three chiral biaryl axes [270]. Shuangancistrotectorines A (**65**), B (**66**), and D (**68**) had the most potent antiplasmodial activity, with IC_{50} values against the K1 strain of 52, 76, and 85 n*M*, respectively, and corresponding SIs of 110, 100, and 610 for K1 vs. L6 mammalian cells. Shuangancistrotectorine C (**67**) was less potent, with a K1 IC_{50} value of 234 n*M* and an SI value of 256.

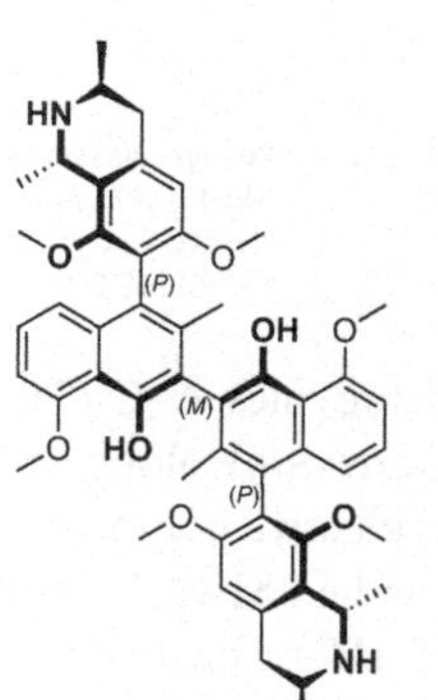

65 (shuangancistrotectorine A)
K1 52 n*M*
SI L6 110

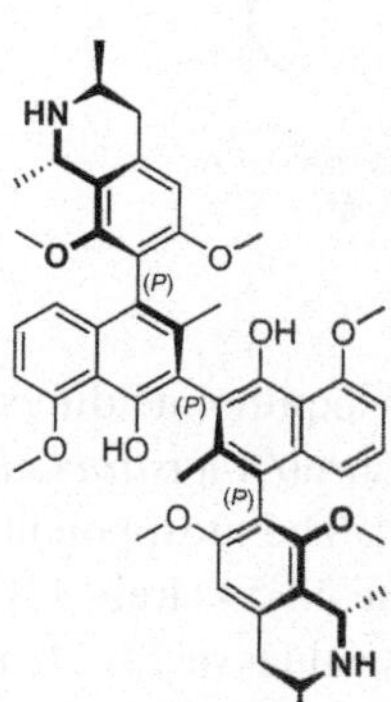

66 (shuangancistrotectorine B)
K1 76 n*M*
SI L6 100

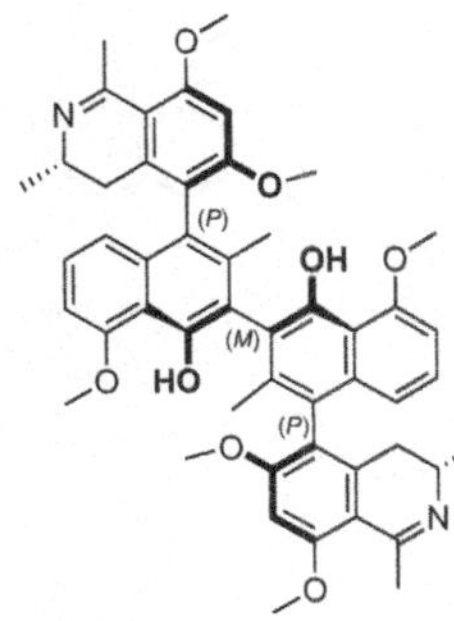

67 (shuangancistrotectorine C)
K1 234 n*M*
SI L6 256

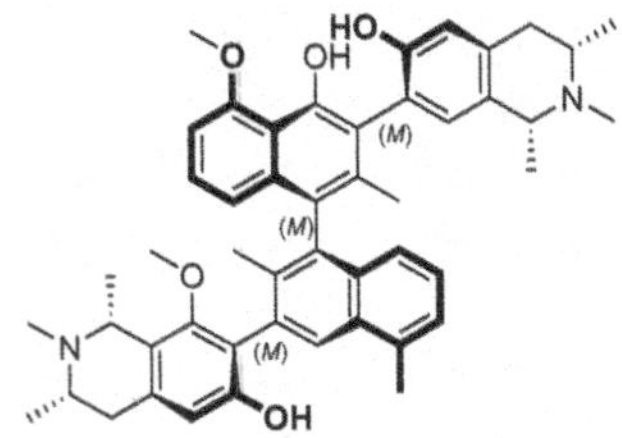

68 (shuangancistrotectorine D)
K1 85 n*M*
SI L6 610

The dimeric naphthylisoquinoline alkaloid mbandakamine A (**69**) was isolated from a Congolese *Ancistrocladus* sp. [271], and both mbandakamines A and C (**70**) were purified from the Central African liana *Ancistrocladus ealaensis* [272]. These two alkaloids have good antiplasmodial activities; mbandakamine A (**69**) has an IC_{50} against the NF54 strain of 60 n*M*, with an SI of 277 for NF54 vs. L6 cells [271], while mbandakamine C (**70**) has an IC_{50} of 47 n*M* against the NF54 strain and 64 n*M* against the K1 strain, with corresponding SIs of 339 and 249 versus L6 cells [272].

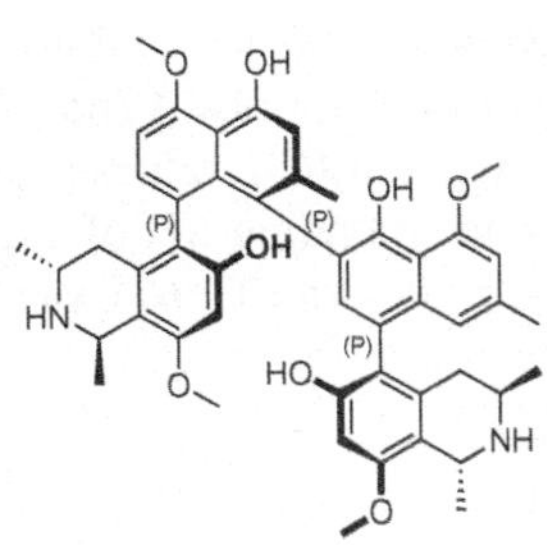

69 (mbandakamine A)
NF54 60 n*M*
SI (L6) 277

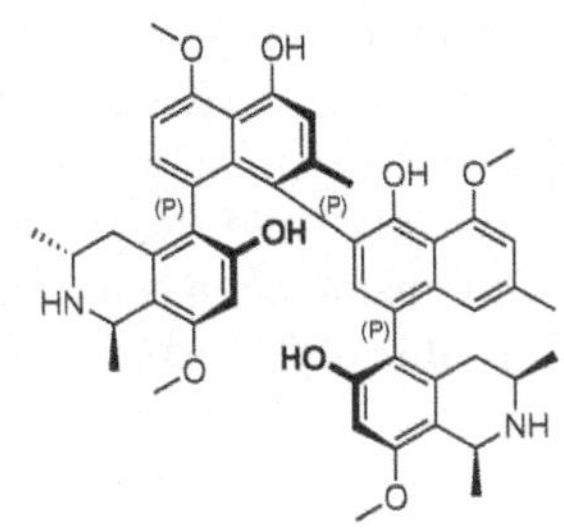

70 (mbandakamine C)
NF54 0.047 μ*M*
K1 0.064 μ*M*
SI (L6) 339/249

The three naphthylisoquinoline dimers jozibrevines A–C (**71–73**) were isolated from the West African shrub *Ancistrocladus abbreviatus*, along with the known dimer jozimine A_2 (**74**) [273]. The most potent and selective compound was jozimine A_2, with an IC_{50} against the NF54 strain 1.4 n*M* and an SI vs. L6 cells of 11,400. Jozibrevines A–C had IC_{50} values of 12, 21, and 55 n*M*, respectively, with corresponding SIs of 660, 250, and 42 against L6 cells [274].

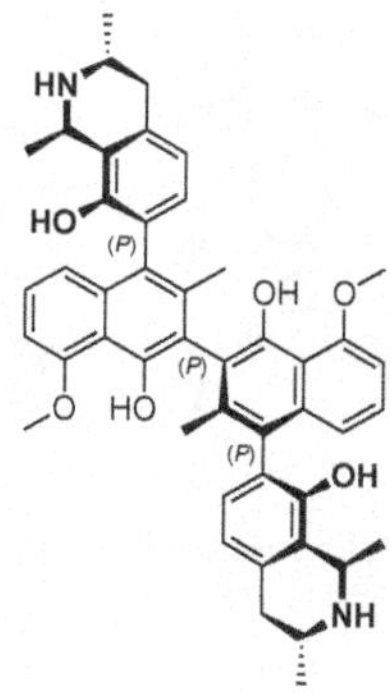

71 (jozibrevine A)
NF54 12 n*M*
SI L6 660

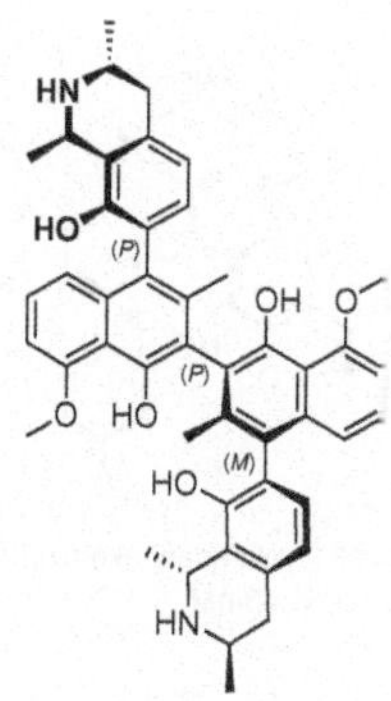

72 (jozibrevine B)
NF54 21 n*M*
SI L6 254

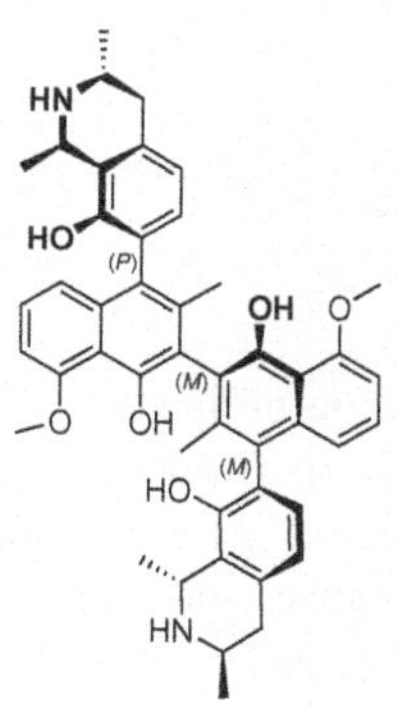

73 (jozibrevine C)
Nf54 55 n*M*
SI L6 42

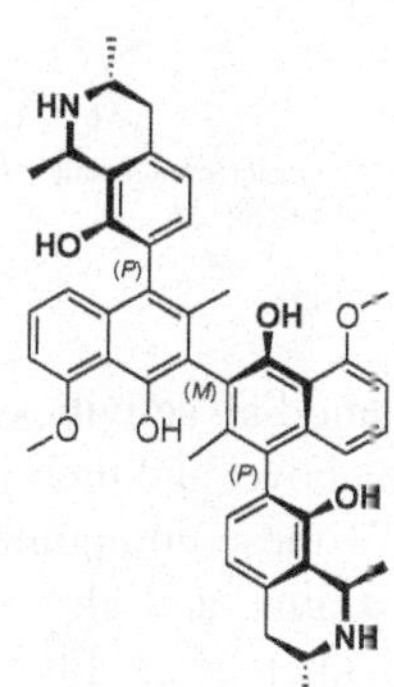

74 (jozimine A_2)
NF54 1.4 n*M*
SI L6 11,400

The naphthylisoquinoline dimers cyclombandakamines A_1 (**75**) and A_2 (**76**) were isolated from a Congolese *Ancistrocladus* sp., and were found to have an oxygen bridge in addition to the usual bi-aryl linkages [275]. They had IC_{50} values against the NF54 strain of 43 and 55 n*M*, respectively, but were not evaluated for mammalian cytotoxicity. Cyclombandakamine A_9 (**77**), spirombandakamine A_3 (**78**) and the previously known mbandakamine C (**70**) were reported from an as yet unidentified *Ancistrocladus* plant from the Democratic Republic of the Congo, offering further evidence of the antiplasmodial activity of this class of compounds [276]. Cyclombandakamine A_9 (**77**) and spirombandakamine A_3 (**78**) had IC_{50} values of 152 and 71 n*M* against *P. falciparum* NF54, with SIs of 127 and 272, using L6 cells [276].

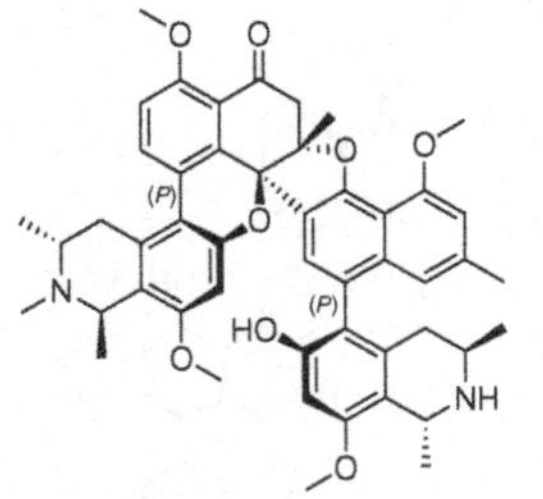

75 (cyclombandakamine A_1)
NF54 43 n*M*

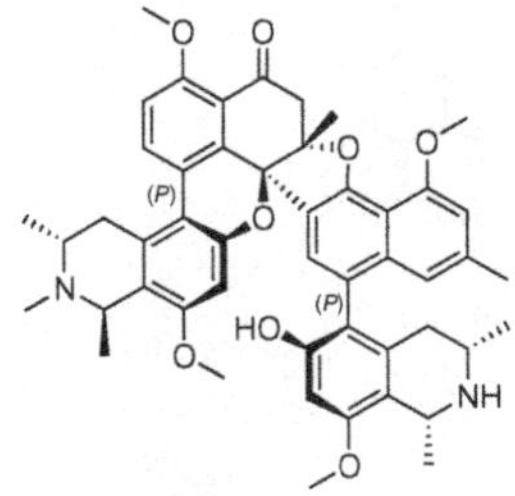

76 (cyclombandakamine A_2)
NF54 55 n*M*

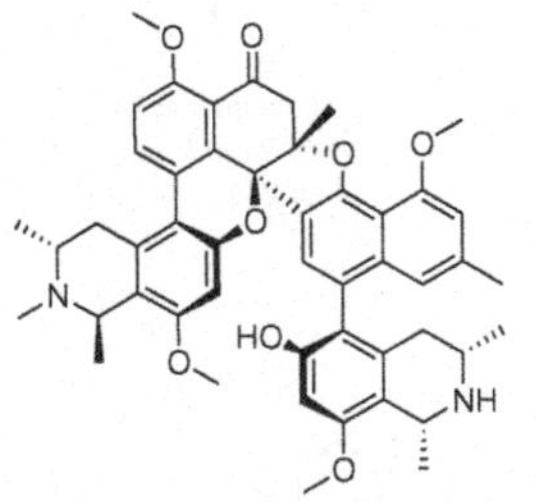

77 (cyclombandakamine A_9)
NF54 152 n*M*
L6 19.3 μ*M*
SI 127

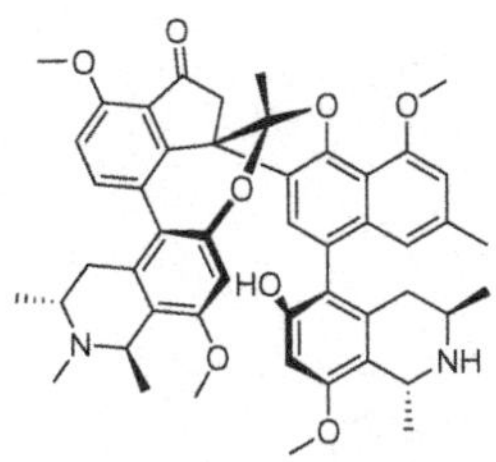

78 (spirombandakamine A_3)
NF54 71 n*M*
L6 19.3 μ*M*
SI 272

The antiplasmodial activities of 30 naphthylisoquinoline alkaloids, including those discussed above, and their synthetic analogs were evaluated in a recent report [277]. Fifteen of these compounds retained significant activity against a panel of drug-resistant strains, and also showed activity against both early and late stage *P. falciparum* gametocytes. Five of the compounds were evaluated for water solubility and metabolic stability, and dioncophylline C (**59**) had the best combination of activity, solubility, and metabolic stability.

Benzophenanthridine Alkaloids

The benzophenanthridine alkaloids nitidine (**79**) and fagaronine (**80**) are produced by various species of the plant genera *Zanthoxylum* and *Fagara*. Nitidine (**79**) is thought to be the main active ingredient in several traditional antimalarial remedies, and it has been found to have an IC_{50} value of 0.49 μ*M* against the FcM29 strain of *P. falciparum* [278], and 0.13 μ*M* against the K39 strain [279] but also with a half maximal cytotoxic concentration (CC_{50}) value of 8.16 μ*M* against Vero cells [278]. It was found to bind heme in a 1:1 complex and inhibit β-hematin formation in vitro, and an in vivo study of nitidine and chloroquine in a four-day suppressive test in mice showed that it inhibited parasitemia, but not as well as chloroquine; at doses of 10 mg/kg/day, chloroquine gave 98% inhibition versus 11% for nitidine [278].

The situation is somewhat different for the related compound fagaronine (**80**), which had an IC_{50} value of 0.50 μ*M* against the 3D7 strain of *P. falciparum* in one study [280], but 0.0076 μ*M* in a second study using synthetic material [281]. This study also showed fagaronine (**80**) to be only very weakly toxic to Vero cells, with a CC_{50} value of 21 μ*M*. In a four-day suppressive test in mice at 10 mg/kg/day fagaronine inhibited parasitemia comparably to chloroquine, with 94.8% inhibition for chloroquine versus 91.5% for fagaronine (**80**). Since both nitidine and fagaronine are charged compounds they are soluble in water, which has the potential to make them orally bioavailable.

Dihydronitidine (**81**) is a neutral compound that shows slightly more potent antiplasmodial activity than nitidine, with an IC_{50} value of 25 n*M* against 3D7 parasites compared with 100 n*M* for nitidine [282]. It is, however, a slow-acting compound, suggesting a different mechanism of action from the faster acting nitidine, and this and its lack of water solubility suggest that it will be a less attractive development candidate than nitidine.

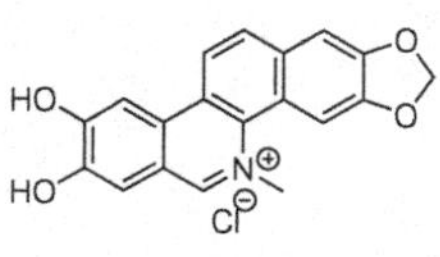

79 (nitidine)
3D7 220 n*M*
FcM29 490 n*M*
K39 130 n*M*
Vero 8.16 μ*M*

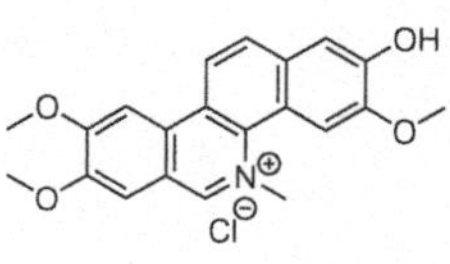

80 (fagaronine)
3D7 50 n*M*

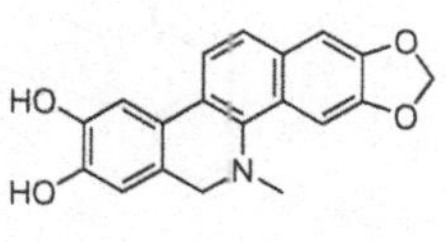

81 (dihydronitidine)
3D7 25 n*M*

Benzylisoquinoline Alkaloids

Benzylisoquinoline alkaloids come in many shapes and sizes, ranging from simple alkaloids such as 10-demethylxylopine (**82**) to oxidized alkaloids such as (–)-milonine (**83**) and dimeric alkaloids such as phaeanthine and cycleanine.

10-Demethylxylopine (**82**) and (–)-milonine (**83**) both show submicromolar activity against *P. falciparum*, with an IC_{50} value of 0.05 μ*M* against the 3D7 strain for **82** [283] and 0.097 μ*M* against the K1 strain for **83** [285]. Mammalian toxicity data are not available for 10-demethylxylopine, but **83** has a CC_{50} value of >200 μ*M* against three mammalian cell lines, giving it a SI of >2000 [284].

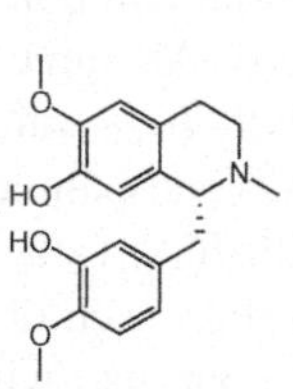

82 (10-demethylxylopinine)
3D7 0.05 μ*M*

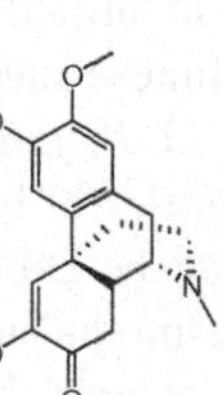

83 ((–)-milonine)
K1 0.097 μ*M*
SI>2000

The bis-benzylisoquinoline alkaloid *O,O*-dimethylgrisabine (**84**) was isolated from the bark of the Malaysian tree *Dehaasia longipedicellata* [284], and had IC_{50} values of 0.031 μM against *P. falciparum* K1 and >200 against A549 human lung cancer cells, for an SI of >6000. The related alkaloids pycmanilline (**85**) and its cyclized analog phaeanthine (**86**) have moderate activity against the K1 strain, with IC_{50} values of 1.6 and 0.8 μM, respectively, but more potent activities of 0.05 and 0.03 μM against the NF54 strain and SIs of 832 and 373 using NF54/MRC-5 (human fibroblast cells) comparison [285]. The dimeric compound cycleanine (**87**) gave an IC_{50} value of 0.08 μM against the 3D7 strain, but no mammalian toxicity data are available [283]. Of these four compounds *O,O*-dimethylgrisabine (**84**) has the greatest potential as an interesting lead compound, provided that it has significant activity in vivo.

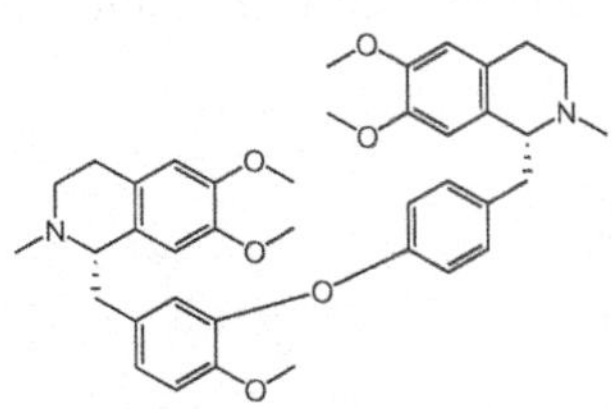

84 (*O,O*-dimethylgrisabine)
K1 0.031 μM
A549 SI >6400

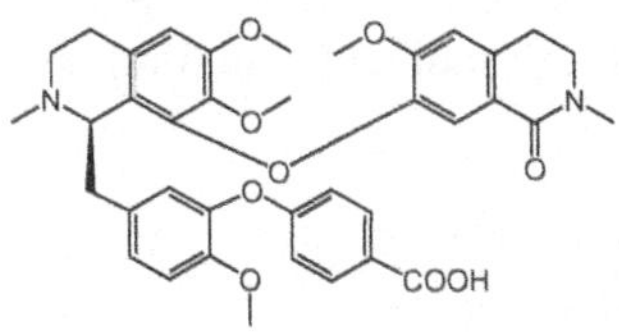

85 (pycmanilline)
K1 1.6 μM
NF54 0.05 μM
MRC-5 41.6 μM
SI 832 (NF54/MRC-5)

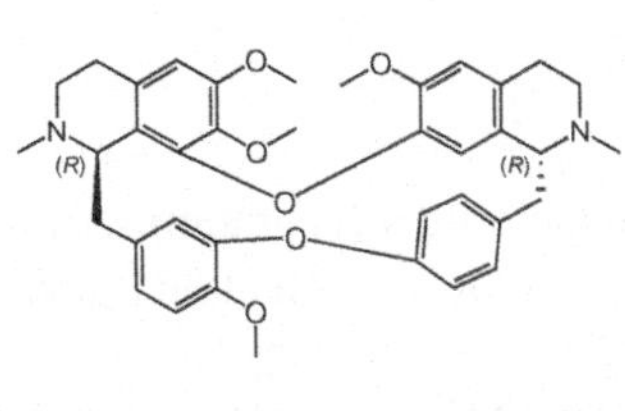

86 (phaeanthine)
K1 0.8 μM
NF54 0.03 μM
MRC-5 11.2 μM
SI 373 (NF54/MRC-5)

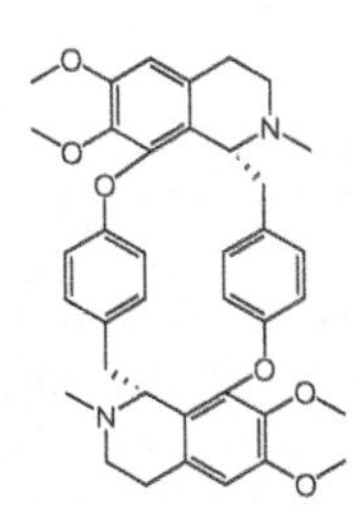

87 (cycleanine)
3D7 0.08 μM

The benzylisoquinoline alkaloids coptisine (**88**), berberine (**89**), and palmatine (**90**) chlorides, were all identified as antimalarial components of a natural products library of traditional Chinese medicines [286]. All three were comparably potent against the 3D7 strain of *P. falciparum*, with IC_{50} values of 0.035, 0.036, and 0.026 μM, and the chlorides of berberine (**89**) and palmatine (**90**) retained submicromolar activity against the Dd2 strain, with IC_{50} values of 0.054 and 0.81 μM, respectively. The antiplasmodial activity of berberine chloride (**89**) has been reported previously as 0.97 μM against the K1 strain but also with the significant activity of 7.3 μM against mammalian cells, so its potential as an antimalarial agent is limited [287]. Coptisine chloride (**88**) was identified as the major antiplasmodial component of *Coptis* rhizome from a collection of 120 traditional Japanese herbal medicines [288]. In this study, it had IC_{50} values of 1.1 μ*M* against the 3D7 strain and 3.1 μ*M* against the Dd2 strain, with SIs of 37.8 and 13.2; the difference in its IC_{50} values from the previous study may be due in part to different incubation times of 48 and 72 h. The corresponding IC_{50} values for **89** were 2.6 and 6.3 μ*M* (SI 3.3 and 1.3) and for **90** were 6.0 and 11.8 μ*M*, with SI values of >16.7 and >8.5. In an in vivo experiment with mice infected with *P. yoelli*, coptisine chloride (**88**) suppressed parasitemia by >80% after 6 days at a dose of 30 mg/kg/day. These alkaloids thus have limited potential as antimalarial agents.

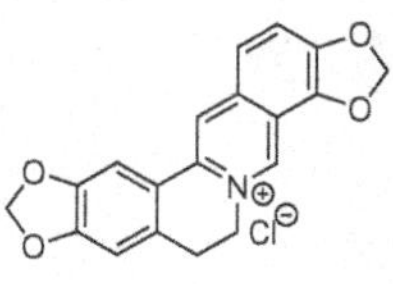

88 (coptisine chloride)
3D7 0.035 μ*M*

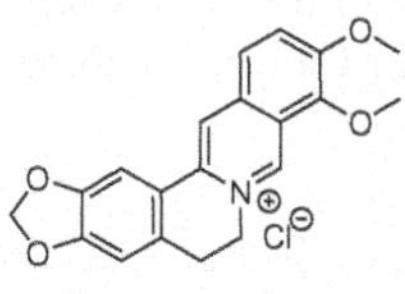

89 (berberine chloride)
3D7 0.036 μ*M*
Dd2 0.054 μ*M*

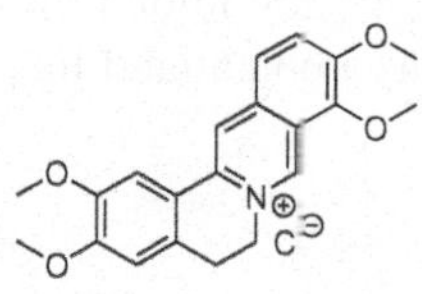

90 (palmatine chloride)
3D7 0.026 μ*M*
Dd2 0.81 μ*M*

Indole and Bisindole Alkaloids

Indole alkaloids are one of the largest classes of alkaloids, with structures ranging from simple indole derivatives to complex terpenoid bisindole alkaloids. For the purposes of this contribution, an alkaloid is classified as an indole alkaloid if it contains an indole ring system, even if it can also be considered as a member of another class of alkaloids. The antimalarial activity of indole alkaloids was reviewed in 2008 [289], and a general review of this class highlighting new structures and biological activities appeared in 2020 [290]. Only a relatively few such alkaloids showed significant activity, as discussed below.

Ellipticine (**91**) and olivacine (**92**) both have micromolar activity against the K1 and 3D7 strains of *P. falciparum*, with IC_{50} values of 0.81 and 0.5 μ*M*, respectively, for (**91**) and 1.4 and 1.2 μ*M* for (**92**). Both were only very weakly cytotoxic to murine macrophages, giving them SIs of >500 for ellipticine and >290 for olivacine. Oral treatment of mice infected with *P. berghei* with ellipticine at 50 mg/kg/day for four days gave 100% survival after 40 days, equivalent to chloroquine (**14**) at 10 mg/kg/day, but similar treatment with olivacine (**92**) only produced an average survival time of 23 days [291].

Cryptolepine (**93**) is the main constituent of the West African climbing shrub *Cryptolepis sanguinolenta*, the roots of which are used for the treatment of malaria in that region. It has been investigated widely as an antimalarial agent, and many analogs have been prepared. A full account of these derivatives is beyond the scope of this review, but some halogenated derivatives such as 7-bromo-2-chlorocryptolepine

(**94**) were significantly more potent than cryptolepine (**93**) itself, with IC_{50} values for **93** and 7-bromo-2-chlorocryptolepine of 0.44 and 0.030 μM, respectively, against the K1 strain of *P. falciparum.* However, neither compound had a good selectivity for K1 versus mammalian MAC 15A cells, with SIs of 9 and 58, respectively [292]. In an evaluation against *P. berghei* in mice with treatment for four days, cryptolepine (**93**) was toxic after two doses at 20 mg/kg/day while 7-bromo-2-chlorocryptolepine at 25 mg/kg/day gave a mean survival time of 15.4 days versus 10 days for untreated controls [292, 293]. It does have high gametocytocidal properties, so its potent antimalarial effect may be due in part to this activity. Cryptolepine (**93**) itself is thus not a useful antimalarial agent, but it is possible that one or more analogs might prove of use, especially since it and its analogs appear to operate by a new mechanism in addition to inhibition of β-hematin formation [292]. Amodiaquine (**15**) has been proposed as a potential drug partner for cryptolepine (**93**) [294].

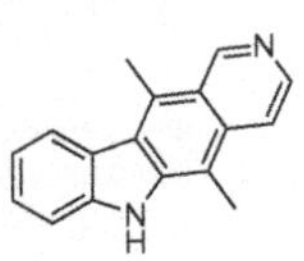

91 (ellipticine)
3D7 0.35 μM
K1 0.81 μM
SI >500

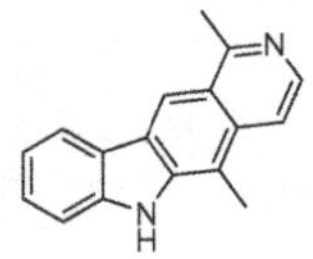

92 (olivacine)
3D7 1.2 μM
K1 1.4 μM
SI >290

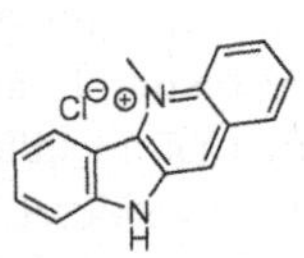

93 (cryptolepine)
K1 0.44 μM
SI 9

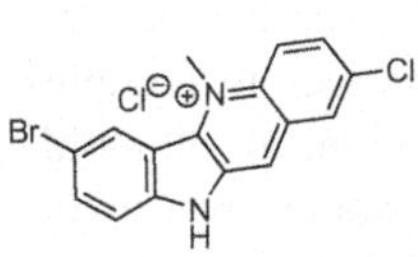

94 (7-bromo-2-chloro-cryptolepine)
K1 0.03 μM
SI 58

Calothrixins A (**95**) and B (**96**) are indolo[3,2-*j*]phenanthridine alkaloids isolated from a *Calothrix* sp. cyanobacterium. They have IC_{50} values of 58 ± 8 nM and 180 ± 44 nM, respectively, against the FAF6 strain of *P. falciparum*, comparable to chloroquine (**14**) with a value of 83 ± 17 nM. Their CC_{50} values against HeLa cells were 40 ± 9 and 350 ± 82 nM, respectively, so their potential as antimalarial agents is limited by their poor antiplasmodial selectivity [295].

Nostocarboline (**97**) is a quaternary β-carbolinium alkaloid isolated from the cyanobacterium *Nostoc* 78-12A [296]. It has an IC_{50} value of 0.19 μM against the K1 strain of *P. falciparum*, and good selectivity with an CC_{50} value of 121 μM against rat myoblast L6 cells for a selectivity ratio of 630 [297].

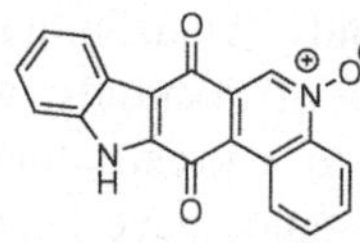

95 (calothrixin A)
FAF6 58 n*M*
HeLa 40 n*M*
SI 0.7

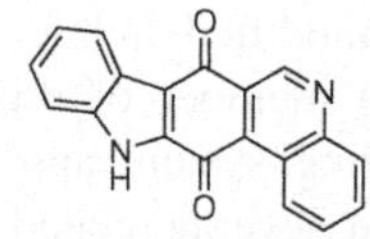

96 (calothrixin B)
FAF6 180 n*M*
HeLa 350 n*M*
SI 1.9

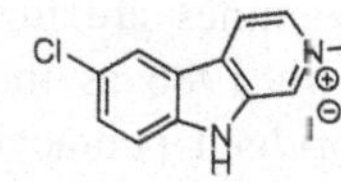

97 (nostocarboline)
K1 190 n*M*
L6 121 μ*M*
SI 630

Violacein (**98**) is a violet bacterial indolocarbazole pigment that is available in low yield from *Chromobacterium* sp., but has been made available from *Escherichia coli* by synthetic biology [298], and this technology has been extended to the generation of violacein analogs [299]. Violacein is active against *P. falciparum* 3D7 with an IC_{50} value of 0.51 μ*M*, and also against both artemisinin-sensitive (ANL1) and artemisinin-resistant (APL5G) patient isolates of *P. falciparum* derived from the Greater Mekong subregion, with IC_{50} values of 0.74 and 0.76 μ*M*, respectively [300]. Its biosynthetically produced 7-chloro analog (**98a**) was slightly more potent, with an IC_{50} value of 0.42 μ*M* against the 3D7 strain, as opposed to 0.51 μ*M* for **98** itself. In addition, **98** appears to have a new mechanism, in that it affects the parasite actin cytoskeleton, suggesting the potential utility of this drug as an antimalarial agent [299].

Dimethylisoborravereine (**99**) and finderole B (**100**) were isolated from the Australian plant *Flindersia acuminata*, and were found to have IC_{50} values of 0.08 and 0.15 μ*M*, respectively, against the Dd2 strain of *P. falciparum*, with SIs of 51 and 14 for the Dd2/HEK293 cell ratio [300]. The case is similar for the related alkaloids borreverine (**101**) and pimentelamine (**102**) [301]. These are thus interesting compounds but probably not really promising antimalarial leads.

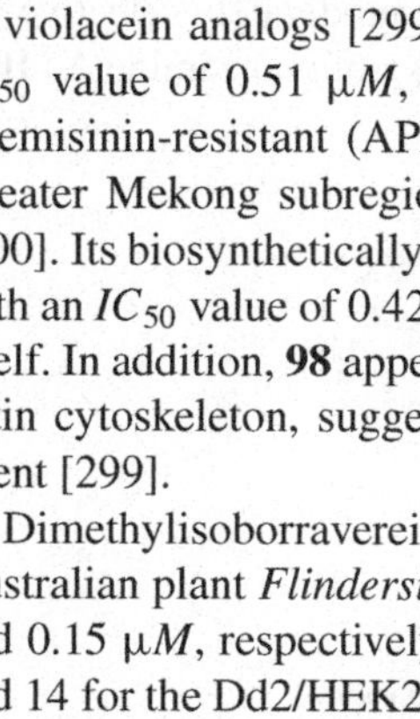

98 R = H (violacein)
3D7 0.51 μ*M*
ANL1 0.74 μ*M*
APL5G 0.76 μ*M*
98a R = Cl (7-chloroviolacein)
3D7 0.42 μ*M*

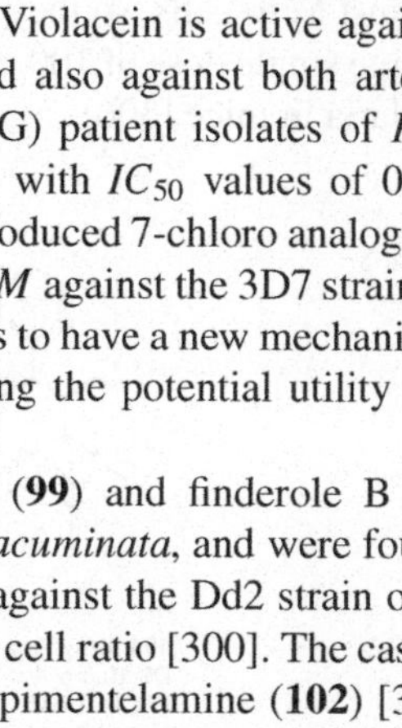

99 (dimethylisoborreverine)
Dd2 0.08 μ*M*
HEK 4.09 μ*M*
SI 51 Dd2/HEK293

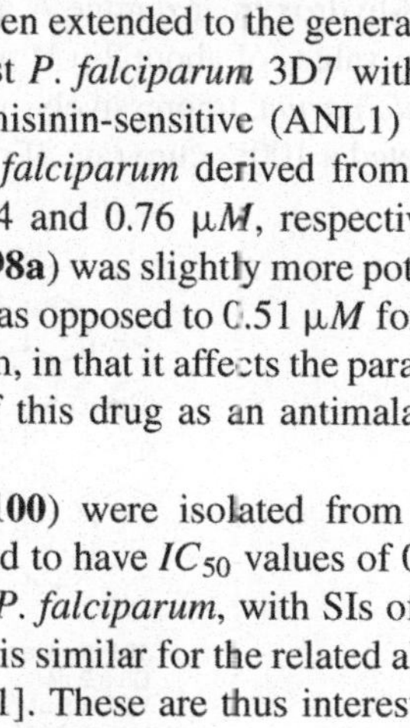

100 (flinderole B)
Dd2 0.15 μ*M*
HEK-293 4.1 μ*M*
SI 27 Dd2/HEK293

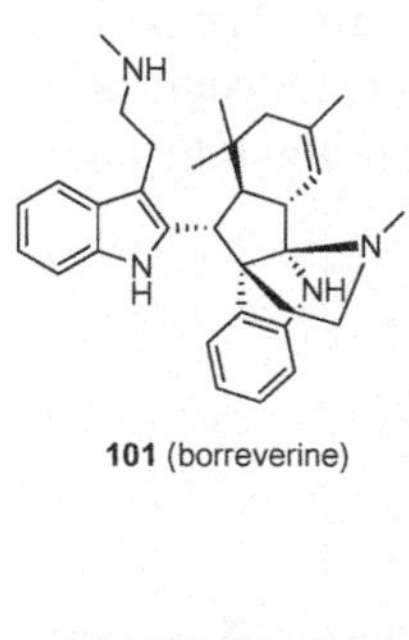

101 (borreverine)

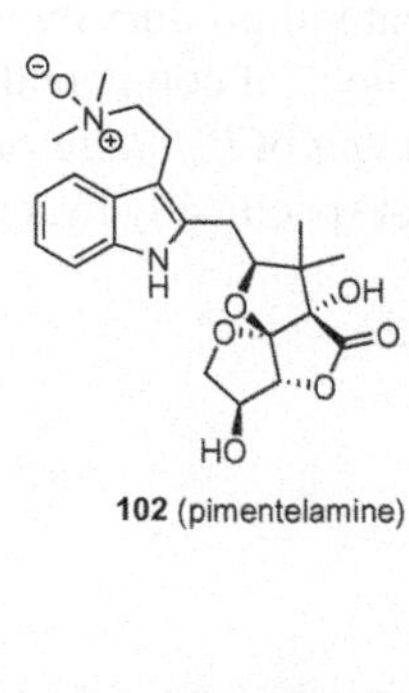

102 (pimentelamine)

The manzamines are highly modified indole alkaloids. Manzamine A (**103**) was isolated in 1986 as the first member of this class of alkaloids, which has a complex bridged pentacyclic ring system appended to a β-carboline alkaloid [302]. Although this alkaloid and several related compounds have been isolated from various marine organisms [303, 304] they have been shown to be produced by the sponge-associated bacterium *Micromonospora* sp. [305]. Both **103** and its congener 8-hydroxymanzamine A (**104**) have promising antimalarial activity, with IC_{50} values for manzamine A of 8.2 nM against the D6 strain of *P. falciparum* and 14.5 nM against the chloroquine-resistant W2 strain. The corresponding figures for 8-hydroxymanzamine A are 10.6 and 14.1 nM, respectively. Both alkaloids had CC_{50} values of about 2 μM against Vero cells, so they are moderately selective for *P. falciparum*. Impressively, a large single dose of 100 mg/kg of manzamine A (**103**) effected a 100% cure rate of malaria in mice [306].

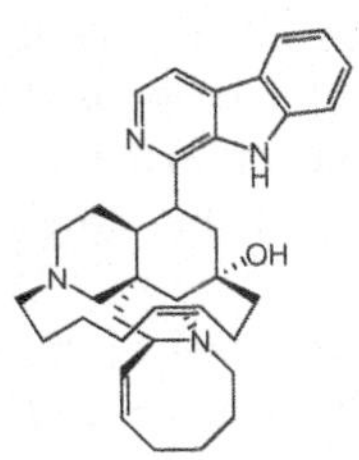

103 (manzamine A)
D6 8.2 n*M*
W2 14.5 n*M*
Vero 2.2 μ*M*

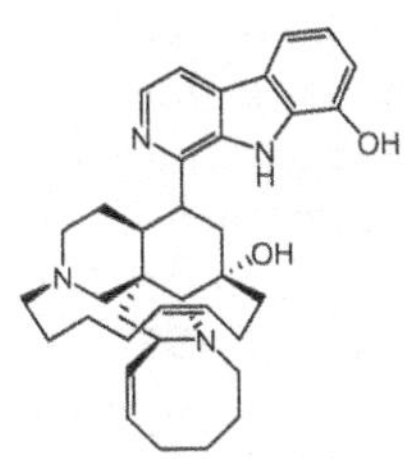

104 (8-hydroxymanzamine A)
D6 10.6 n*M*
W2 14.1 n*M*
Vero 1.9 μ*M*

The bisindole alkaloid ochrolifuanine A (**105**) is an example of an alkaloid with submicromolar potency against three strains of *P. falciparum*, but with less than excellent selectivity for plasmodial cells compared with HCT-116 colon cancer cells [307]. It and related natural products did, however, spur the exploration of simple synthetic β-carbolines as antimalarial agents, and three of these with interesting origins or activities are noted here.

Cipargamin (NITD609 or KAE609) (**106**) was identified as part of a phenotypic high-throughput screening of ~12,000 natural products and natural product-like synthetic compounds. It had subnanomolar activity against the NF54 and K1 *P. falciparum* strains, and was found to inhibit the early and late development of *P. falciparum* gametocytes [308]. A single oral dose of 1 mg/kg gave a 13.5% survival of *P. berghei*-infected mice, superior to artesunate's 7.3% survival, and a single oral dose of 100 mg/kg gave 100% survival [309]. This compound meets the criteria for drug development, and was in Phase II clinical trials as of 2020 [310, 311].

The synthetic natural product mimic MMV008138 (**107**) is a component of the so-called "malaria box" of compounds with confirmed activities against the asexual intraerythrocytic stages of *P. falciparum* and low cytotoxicities [312]. It was found to target the apicoplast specifically by a mechanism similar to that of fosmidomycin by targeting the methylerythritol phosphate pathway for isoprenoid biosynthesis [313] but at a different enzymatic step [314].

The related synthetic β-carboline **108** showed more than 75% inhibition against both the 3D7 and RKL-9 strains of *P. falciparum* at the lowest tested concentration (2.8 μ*M*) and gave a chemosuppression of 91% in *P. berghei*-infected mice at a dose of 100 mg/kg [315]. This class of alkaloids and related synthetic compounds thus appears to have good antimalarial potential.

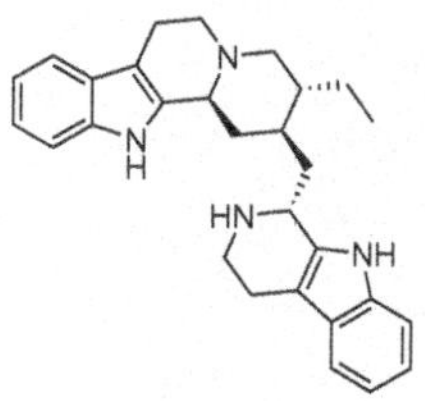

105 (ochrolifuanine A)
FCA 20 0.12 μ*M*
FCB1-R 0.27 μ*M*
W2 0.49 μ*M*
HCT-116 16.1 μ*M*

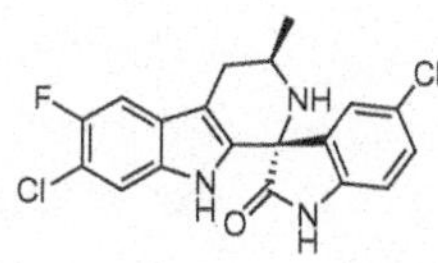

106 (cipargamin,
NITD609, KAE609)
NF54 0.5 n*M*
K1 0.6 n*M*

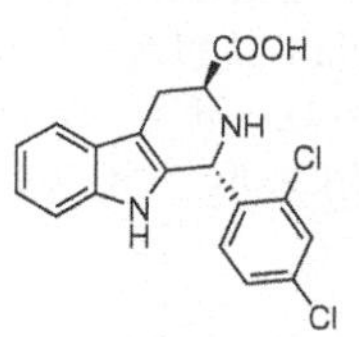

107 ((1*R*,3*S*)-MMV008138)

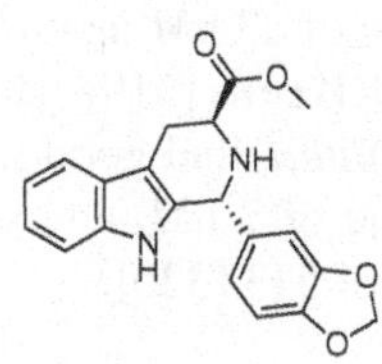

108

Terpenoid indole alkaloids are a major subclass of indole alkaloids, and have been well investigated for antimalarial activity. As one example, Frédérich et al. evaluated 69 indole and bis-indole alkaloids from *Strychnos* spp. for antiplasmodial activity. Eleven alkaloids had submicromolar IC_{50} values against the CQ-resistant W2 strain of *P. falciparum*, but none of them had adequate SIs against the HCT-116 mammalian cell line in order to be considered as good hit compounds [307].

Simple terpenoid indole alkaloids have not usually shown strong selective antiplasmodial activity. Thus, aspidoscarpine (**109**) had an IC_{50} value of 15 μ*M* against the W2 strain of *P. falciparum*, and a SI of only 55 for W2 vs. HepG2 cells [316]. The quaternary indole alkaloid nicalaterine A (**110**) was more potent, with a 110 n*M* IC_{50} against the D7 strain and an SI of >450 against the HL-60 leukemia cell line [317]. In some cases, chemical modification can improve the activity of these alkaloids; as an example, Paciaroni et al. prepared 14 derivatives of the weakly active yohimbine (IC_{50} > 50 μ*M* to the Dd2 strain) and found one designated Y7j (**111**) that showed an IC_{50} value of 0.33 μ*M* against the same strain, and a SI of over 100. It was found to be a fast-acting parasiticidal agent by a series of kill kinetic experiments [318].

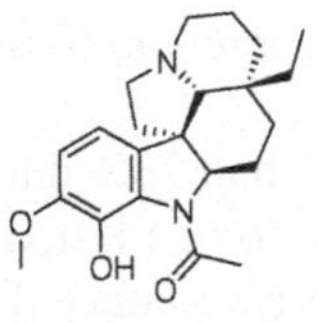

109 (aspidoscarpine)
W2 15 μ*M*
HepG2 813 μ*M*
SI 55

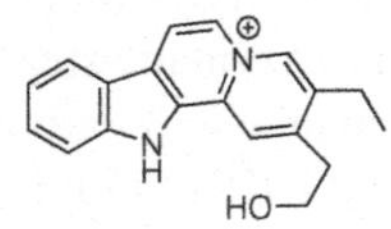

110 (nicalaterine A)
3D7 110 n*M*
HL-60 >50 μ*M*
SI >450

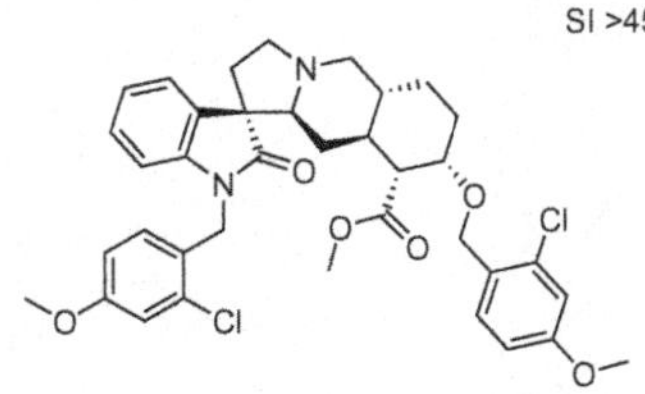

111 (Y7j)
Dd2 0.33 μ*M*
SI >100

The unusual bisindole alkaloid 3′,4′-dihydrousambarensine (**112**) is one of the relatively few such alkaloids with good potency and selectivity against *P. falciparum*, with an IC_{50} value of 23 n*M* against the W2 strain and a SI of 1.475 in comparison with mammalian KB cells [319]. The bisindole alkaloid **113**, isolated from the leaves of *Leuconotis griffithii*, had good potency against the 3D7 strain of *P. falciparum*, with an IC_{50} value of 7 n*M*, and only moderate cytotoxicity against HL-60 cells, giving a SI value of 543 [320].

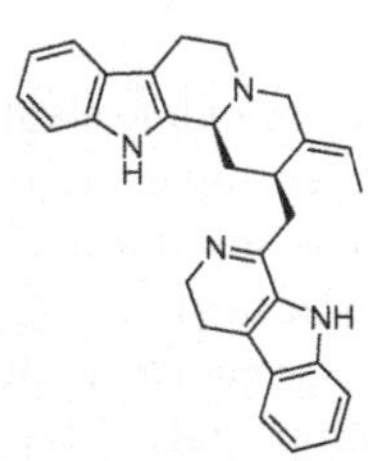

112 (3′,4′-dihydrousambarensine)
W2 23 n*M*
KB 33.9 μ*M*
SI 1475

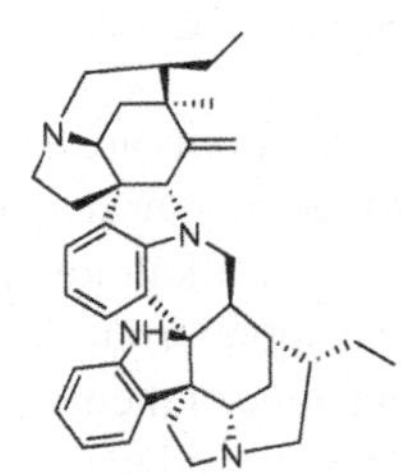

113
3D7 7 n*M*
HL-60 3.8 μ*M*
SI 543

Bisnicalaterine C (**114**) was the most potent of a series of related alkaloids isolated from *Hunteria zeylanica*, with an IC_{50} of 0.05 μ*M* against the 3D7 strain and a CC_{50} of >50 μ*M* against the HL-60 mammalian cell line, for a SI of >1000. As noted above, the simpler nicalaterine A was almost half as potent, suggesting that the important part of the pharmacophore is the ionic indole portion of the structure [317]. Strychnogucine B (**115**) was isolated from *Strychnos icaja*, and a sample for in vivo evaluation was prepared by semisynthesis from strychnine. It had only moderate antiplasmodial activity against the FCA 20 and FC B1-R strains, but gave an IC_{50} of 0.085 μ*M* against the W2 strain [321]. An in vivo study by Beaufay et al. [322] using the 4-day suppressive test on mice infected with *P. berghei* at 30 mg/kg/day showed that parasitemia was reduced by 35.8% on day 5 and by 60.3% on day 7, and survival was increased by at least 24 h for 80% of the mice. However, only 50% of the mice on strychnogucine B survived to day 10, while 100% of the mice on chloroquine survive to this time. Strychnogucine B (**115**) is thus an interesting compound, but probably not suitable for further development.

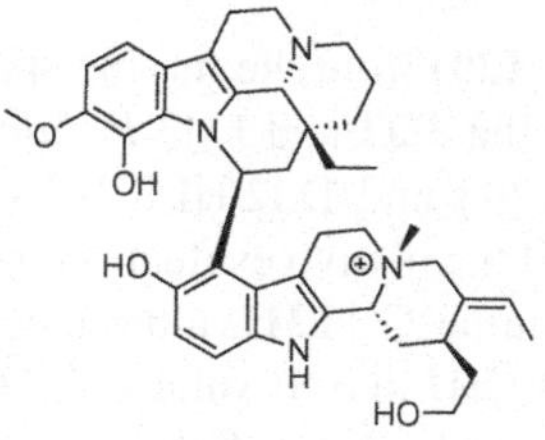

114 (bisnicalaterine C)
3D7 50 *nM*
HL-60 >50 μ*M*
SI >1000

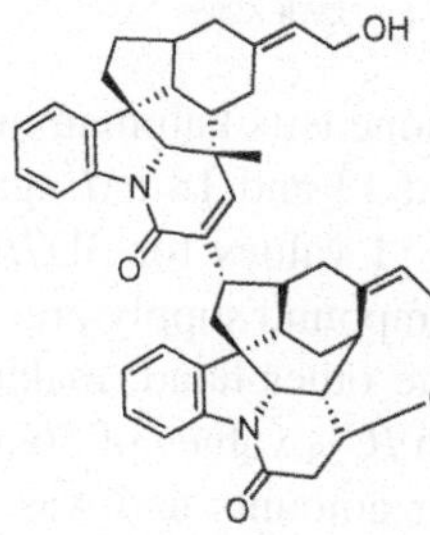

115 (strychnogucine B)
W2 0.085 μ*M*

Quinone and Iminoquinone Alkaloids

Several quinone or iminoquinone alkaloids have significant antiplasmodial activity. The thiaplakortones A–D (**116–119**) are examples of quinonoid alkaloids with significant in vitro antiplasmodial activity and reasonable selectivity as compared with cytotoxicity to HEK293 cells. The most potent compound is thiaplakortone A, with an IC_{50} of 0.051 μ*M* against the 3D7 strain and 0.007 μ*M* against the Dd2 strain of *P. falciparum.* It was also the most cytotoxic to HEK293 cells, so its SI was only 78 for the 3D7/HEK293 comparison and 591 for the Dd2/HEK293 comparison. Thiaplakortone D (**119**) although less potent to the *P. falciparum* strains, was also less toxic to HEK293 cells, and thus had an improved SI. Although these alkaloids are not potent enough for advancement to in vivo studies, it is possible that medicinal chemistry input could generate improved lead compounds [323].

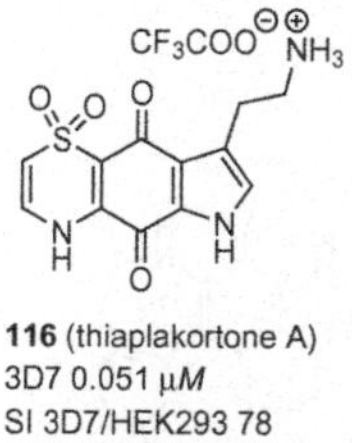

116 (thiaplakortone A)
3D7 0.051 μ*M*
SI 3D7/HEK293 78
Dd2 0.007 μ*M*
SI Dd2/HEK293 591

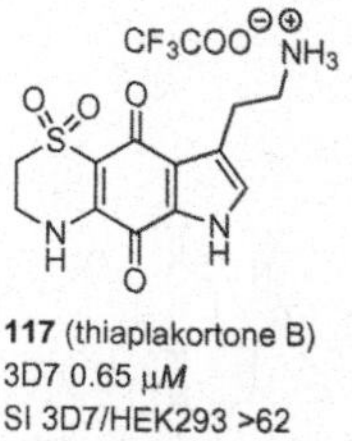

117 (thiaplakortone B)
3D7 0.65 μ*M*
SI 3D7/HEK293 >62
Dd2 0.092 μ*M*
SI Dd2/HEK293 >435

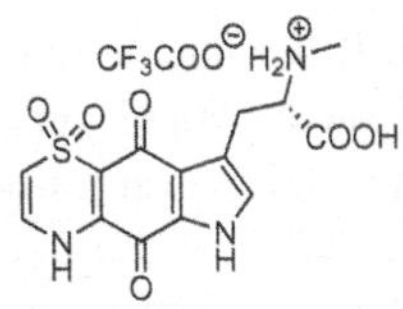

118 (thiaplakortone C)
3D7 0.31 μ*M*
SI 3D7/HEK293 >129
Dd2 0.17 μ*M*
SI Dd2/HEK293 >223

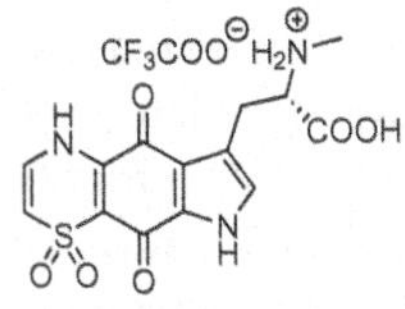

119 (thiaplakortone D)
3D7 0.28 μ*M*
SI 3D7/HEK293 >285
Dd2 0.16 μ*M*
SI Dd2/HEK293 SI >500

The iminoquinone tsitsikammamine C (**120**) from the marine sponge *Zyzzya* sp. had IC_{50} values of 13 and 18 n*M* against the 3D7 and Dd2 *P. falciparum* strains, respectively, and SI values for 3D7/HEK293 and Dd2/HEK293 of 276 and 200 [324]. Limited compound supply precluded the in vivo evaluation of this interesting compound. On the other hand, makaluvamine G (**121**), although less potent and less selective with IC_{50} values of 36 and 39 n*M* and SI values of 276 and 200, was available in larger amounts and was evaluated against *P. berghei* in mice using a standard 4-day test at a dose of 8 mg/kg/day. It was found to suppress parasitemia by 48% on day four compared with a control, so it is interesting to speculate on the potential in vivo activity of **120**. The simpler compound damirone A (**122**) was also less potent, with IC_{50} values of 1880 and 360 n*M*, but improved SI values of >63 and >333 against HEK293 human embryonic kidney cells [324]. These studies suggest that medicinal chemistry studies of this class of compounds is likely to be fruitful. The semisynthetic discorhabdin derivative (**123**) exhibited pronounced activity toward the K1 strain (IC_{50} value of 80 n*M*) with an excellent SI of 512 versus the L6 rat skeletal myoblast cell line, and it is thus a promising lead compound [325].

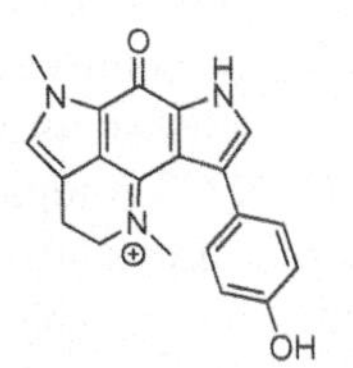

120 (tsitsikammamine C)
3D7 13 n*M*
SI 3D7/HEK293 276
Dd2 18 n*M*
SI Dd2/HEK293 200

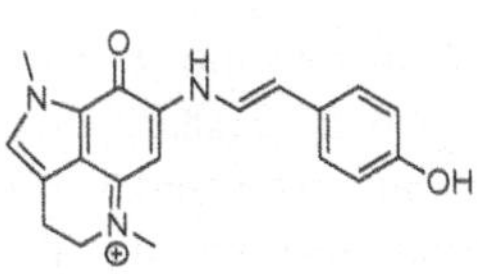

121 (makaluvamine G)
3D7 36 n*M*
SI 3D7/HEK293 55
Dd2 39 n*M*
SI Dd2/HEK293 51

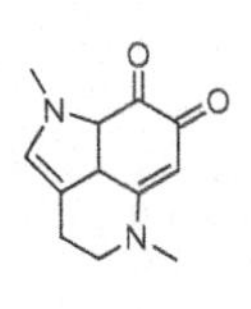

122 (damirone A)
3D7 1.9 μ*M*
SI 3D7/HEK293 >63
Dd2 0.36 μ*M*
SI Dd2/HEK293 >333

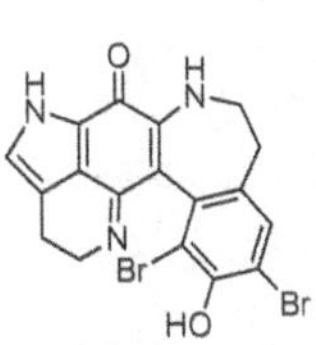

123
K1 80 n*M*
SI K1/L6 51

Miscellaneous Alkaloids

Febrifugine (**124**) was isolated from *Dichroa febrifuga* as the active component against *P. falciparum* [326], and it has been thoroughly investigated as a potential antimalarial compound since 1950 [327]. Its use as an antimalarial agent is precluded by various adverse side effects, in part due to the formation of reactive metabolites such as **125**, but synthetic analogs including **126** [328] and **127** [329] are more potent than febrifugine and have potential as clinical antimalarial agents. In particular, analog **127** had a 50% cure rate at 2 mg/kg/day on the blood stage of *P. falciparum* in Aotus monkeys, and a 100% cure rate at 8 mg/kg/day [329]. Halofuginone (**128**) is the active constituent of the antiprotozoal veterinary drug Stenerol. It is more toxic than febrifugine and has been investigated as an anticancer agent, but it has a future as a potential antimalarial agent [330]. It has been shown to target the cytoplasmic prolyl tRNA synthetase in *P. falciparum* (PfcPRS) [331]. The chemistry and biology of febrifugine (**124**) and halofuginone (**128**) up to 2014 have been reviewed [332]. The analog **129** has also been proposed as an effective antimalarial agent [333]. These and similar compounds thus offer potential for the development of new antimalarial drugs based on febrifugine (**124**).

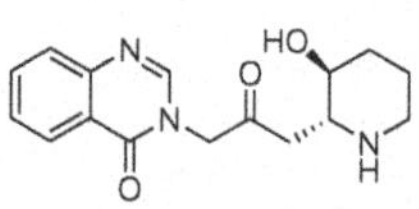
124 (febrifugine)

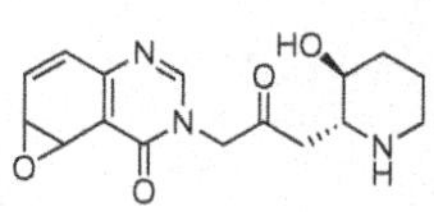
125

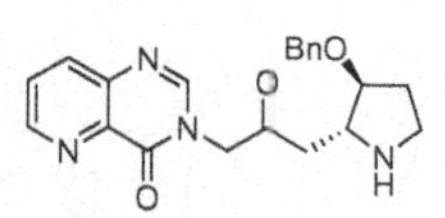
126

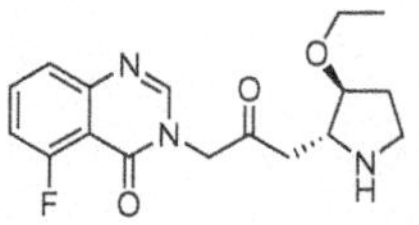
127

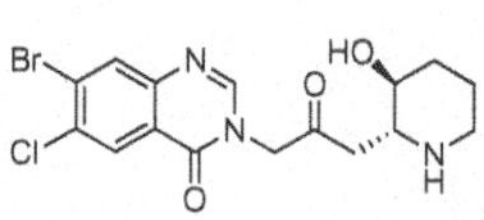
128 (halofuginone)

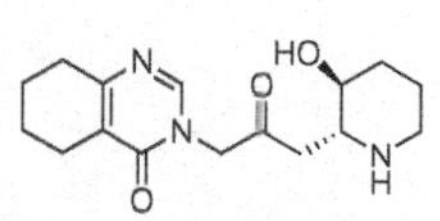
129

Both pactamycin (**130**) and 7-deoxypactamycin (**131**), isolated from *Streptomyces pactum* [334] have potent antiplasmodial activities, with IC_{50} values of 16 and 0.4 n*M*, respectively, against the drug-resistant K1 strain of *P. falciparum*. In addition, 7-deoxypactamycin has an SI of 70 as determined by the ratio of its CC_{50} values against the MRC-5 human diploid embryonic cell line and the K1 strain [335]. Although this is lower than desirable, it is conceivable that synthetic modification could improve selectivity and make it a viable development candidate.

The isolation of the β-lactone salinosporamide A (**132**) from a bacterium of the genus *Salinospora* was reported in 2003 [336], and it was shown to be potently cytotoxic and a proteasome inhibitor. It was later evaluated for antiplasmodial activity and found to inhibit the 3D7 strain of *P. falciparum*, with an IC_{50} value of 11.4 n*M*, and the chloroquine-resistant FCB strain with an IC_{50} of 19.6 n*M*. Its target appears to be the *Plasmodium* proteasome complex. It was also evaluated against the rodent malaria parasite *P. yoellii* in mice, and was found to inhibit parasite growth at the low dose of 130 μg/kg [337]. Although these results are encouraging, the potent cytotoxicity of salinosporamide A is likely to limit its development as an antimalarial agent unless a more selective antimalarial proteasome inhibitor can be developed.

The streptothricin-like antibiotic BY-81 (**133**) (also known as citromycin) was isolated in the late 1960s by Japanese and American groups. It had good activity against both Gram-positive and Gram-negative bacteria [338, 339]. It was found later to have modest activity against the K1 strain of *P. falciparum*, with an IC_{50} value of 0.68 μ*M*. Its cytotoxicity to MRC-5 cells was >200 μ*M*, so it has good selectivity for plasmodial cells but only modest overall activity [335]. It is thus deemed to be an unlikely development candidate for malaria treatment.

The phenanthroindolizidine alkaloid dehydroantofine (**134**) exhibited an IC_{50} value of 0.028 μ*M* against the 3D7 strain and of >56 μ*M* against L929 mouse fibroblast cells, giving it a promising SI of >1964. Its isomer tylophoridicine (**135**), was slightly less potent, with an IC_{50} value of 0.058 μ*M* against the 3D7 strain [340].

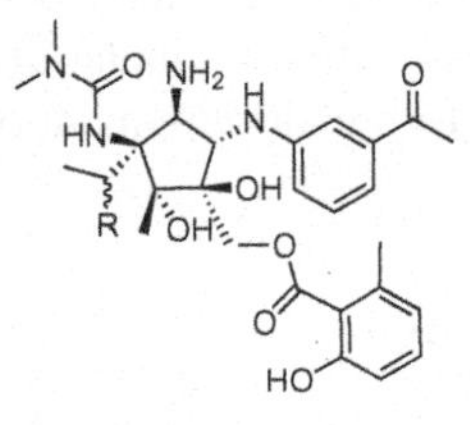

130 R = O (pactamycin)
K1 16 n*M*
131 R = H (7-deoxypactamycin)
K1 0.4 n*M*
SI K1/MRC-5 70

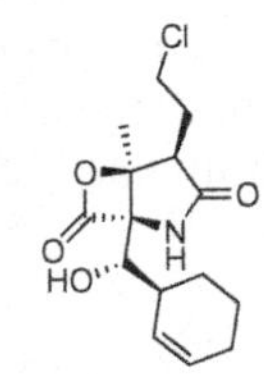

132 (salinosporamide A)
3D7 11.4 n*M*
FCB 19.6 n*M*

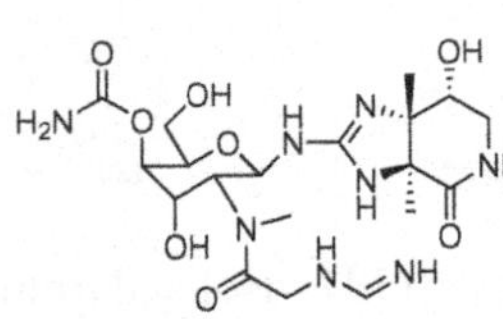

133 (BY-81)
K1 0.68 μ*M*
MRC-5 >200
SI >300

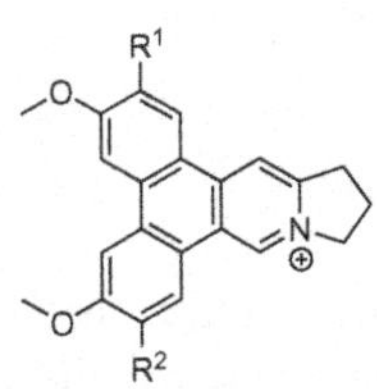

134 R[1] = OMe, R[2] = H
3D7 28 n*M*
135 R[1] = H, R[2] = OMe
D7 0.028 μ*M* SI 1900

Both cycloprodigiosin and metacycloprodigiosin (**136**) are tripyrrole pigments; the latter was first isolated in 1968 from *Streptomyces longisporus ruber* [341]. These two compounds have a good activity against *P. falciparum*, with cycloprodigiosin showing an IC_{50} value of 11 n*M* against the FCR-3 strain, more potent than chloroquine, mefloquine, and artesunate in the same assay [342]. Metacycloprodigiosin (**136**) has good activity against the *P. falciparum* K1 strain with an IC_{50} of 13 n*M*. It is moderately selective for *P. falciparum*, with a CC_{50} of 3.4 μ*M* against Vero cells and a SI of 260 [343]. Isoheptylprodigiosin (**137**) was isolated in 2020 from the marine organism *Vibrio spartinae* with the help of genomics–metabolomics profiling [344], and a recent paper reports its synthesis along with those of 2-(*p*-hydroxybenzyl)-prodigiosin and geometric isomers of tambjamine MYP1. Both isoheptylprodigiosin (**137**) and some related compounds cured erythrocytic *P. yoelii* in mice in a 4-day trial at 25 mg/kg/day [345].

Ingamine A (**138**) is a known piperidine-type alkaloid isolated from the sponge Petrosid Ng5 Sp5, along with its new dihydro congener dihydroingamine A (**139**) and other alkaloids. Dihydroingamine A (**139**) proved to be the most potent antiplasmodial compound, with an IC_{50} of 173 n*M* against the *P. falciparum* D6 and 127 n*M* against *P. falciparum* W2 strain, and with a SI value of over 100 against Vero cells. Ingamine A (**138**) was slightly less potent, with IC_{50} values of 200 and 160 n*M*, respectively [346].

Psammaplysin H (**140**) is a bromotyrosine alkaloid isolated from a marine sponge of the genus *Pseudoceratina.* It had an IC_{50} value of 410 n*M* against *P. falciparum* 3D7, and a SI of >40 against HEK293 mammalian cells [347].

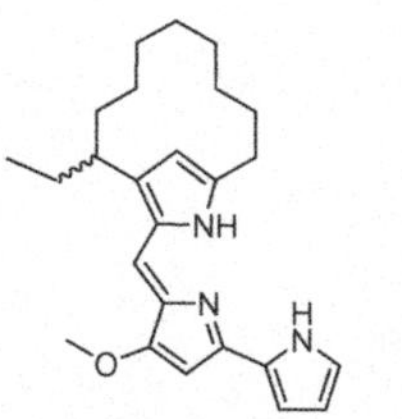

136 (metacycloprodigiosin)
K1 13 n*M*
Vero 3.4 µ*M*
SI 261

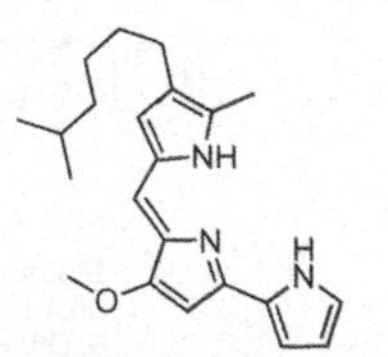

137 (isoheptylprodigiosin)
D6 4.1 n*M*
Dd2 2.1 n*M*
HepG2 1490 n*M*
SI (dd2/HepG2) 360

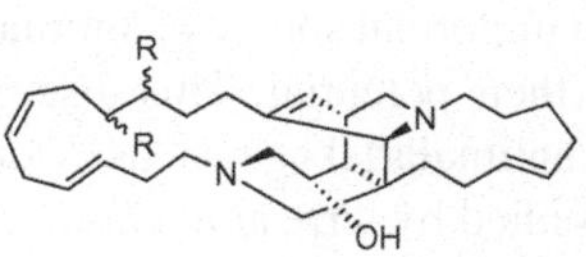

138 R = bond (ingamine A)
D6 200 n*M*
W2 160 n*M*
139 R = H (dihydroingamine A)
D6 173 n*M*
W2 127 n*M*
SI W2/Vero >100

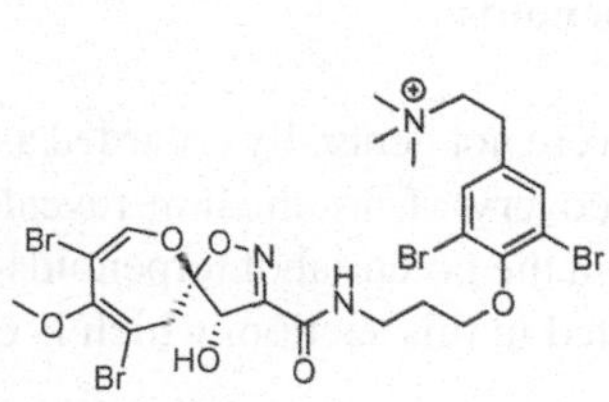

140 (psammaplysin H)
3D7 410 n*M*
SI 3D7/HEK293 >40

The unusual Amaryllidaceae alkaloids cripowellins A (**141**) and B (**142**) were isolated originally from *Crinum powellii* [348], but were then reisolated together with their congeners cripowellins C and D from *Crinum erubescens* [349]. Both cripowellins **141** and **142** had potent antiplasmodial activity, with IC_{50} values of

30 and 180 n*M*, respectively, against the Dd2 strain. However, they also showed potent cytotoxic activity to mammalian A2780 cells, thus limiting their potential as antimalarial drugs [349].

Epoxycytochalasin H (**143**) from the endophytic fungus *Diaporthe miriciae* had submicromolar activity against both the D6 and W2 strains of *P. falciparum*, with IC_{50} values of 0.1 and 0.08 μM, and good selectivity, using Vero cells [350].

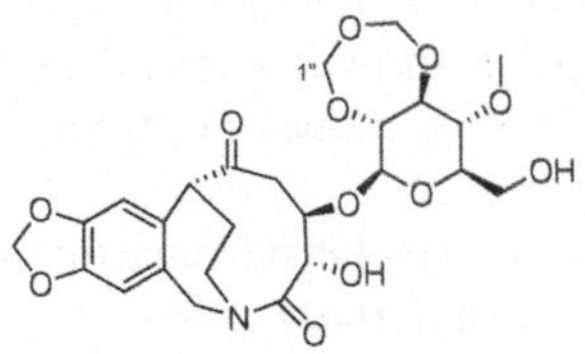

141 (cripowellin A)
Dd2 30 n*M*
SI Dd2/A2780 1

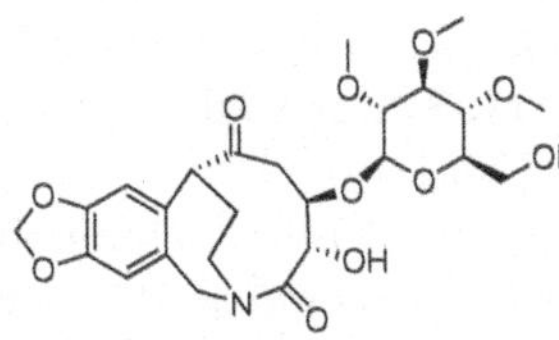

142 (cripowellin B)
Dd2 180 n*M*
SI Dd2/A2780 0.1

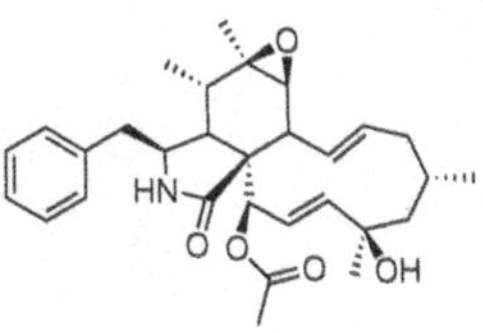

143 (epoxycytochalasin H)
D6 100 n*M*
SI D6/Vero >92
W2 80 n*M*
SI W2/Vero >121

3.5.3 Terpenoids

Terpenoids were not generally regarded as an important source of antimalarial agents until the discovery of artemisinin revealed their potential. This discovery created new interest in the potential of terpenoids as antimalarial agents, as evidenced by the examples listed in this section, which is classified by terpenoid class.

Monoterpenoids

Only a few antimalarial monoterpenoids have been identified as having antiplasmodial activity, and these are generally not highly potent. As one example, a recent review [351] of monoterpenes from Red Sea algae cited only one reference to antiplasmodial activity [352], and the most potent compound of the nine halogenated monoterpenes isolated only had an IC_{50} value of 16 μM against the D10 strain of *P. falciparum*.

Isonitrile-Containing Terpenoids

Terpenoids substituted with one or more isonitrile (isocyano) groups often have strong antiplasmodial activity, so they are considered as a separate group. The literature on marine terpenoids derived from inorganic cyanide through 2014 was the subject of a detailed review by Schnermann and Shenvi [353], with an emphasis on the syntheses of these compounds. As examples, the sesquiterpenoid axisonitrile 3 (**144**) had IC_{50} values of 71 and 614 n*M* against the W2 and D6 strains of *P. falciparum*, while its CC_{50} to KB-3 cells was over 86 μ*M*, so its selectivity is comparable to that of known drugs. The corresponding isothiocyanate derivative was significantly less potent, indicating the importance of the isonitrile function [354]. 2-Isocyanoclovene (**145**), 2-isocyanoclovane (**146**), and 4,5-*epi*-10-isocyanoisodauc-6-ene (**147**) were also each found to have submicrolar activity against the 3D7 and Dd2 strains of *P. falciparum*, with IC_{50} values ranging from 0.3 to 0.87 μ*M*. These compounds showed reasonable parasite-specific selectivity, with 45–158-fold better activity for *P. falciparum* versus mammalian neonatal foreskin fibroblast (NFF) cells [355].

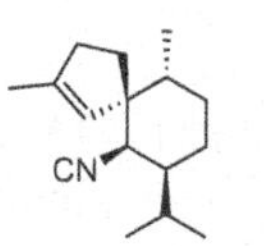

144 (axisonitrile 3)
W2 71 n*M*
D6 614 n*M*
KB >86 μ*M*
SI W2/KB >1200

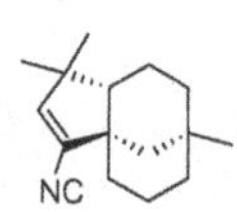

145 (2-isocyanoclovene)
Dd2 0.36 μ*M*
3D7 0.30 μ*M*

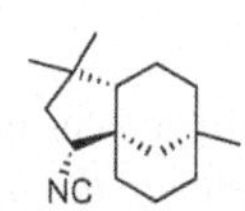

146 (2-isocyanoclovane)
Dd2 0.83 μ*M*
3D7 0.29 μ*M*

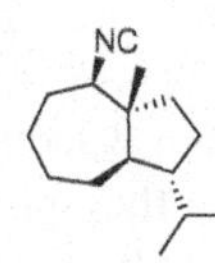

147 (4,5-*epi*-10-isocyano-isodauc-6-ene)
Dd2 0.87 μ*M*
3D7 0.26 μ*M*

Four diterpene isonitriles (**148–151**) were isolated from the tropical marine sponge *Cymbastela hooperi*, and were found to have potent antiplasmodial activities, with the IC_{50} values shown below [356]. (±)-7-Isocyano-11(20),14-epiamphilectadiene (**150**) has been synthesized in about 18% overall yield from an advanced intermediate [357], and (+)-7,20-diisocyanoadociane (**148**) was later synthesized in 17 linear steps by the same group. The availability of synthetic material enabled additional biological testing, and both **148** and **150** were found to have moderate liver-stage activity, with an IC_{50} value of 1.3 and 0.93 μ*M* against *P. berghei* (PbLuc) [358]. The mechanism of action of these and other isonitriles was proposed to be by inhibition of heme detoxification. It was shown that both (+)-7,20-diisocyanoadociane, (**148**) as well as axisonitrile-3 (**144**) interacted with heme by forming a coordination complex with the heme iron [359], although metal binding alone cannot explain the observed activities [353]. The heme detoxification proposal was later supported by the direct use of a β-hematin inhibition assay, which indicated that isonitrile **148** totally inhibited β-hematin crystallization [360].

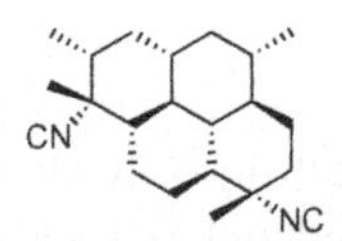

148 (7,20-diisocyanoadociane)
D6 0.015 μ*M*
Dd2 0.016 μ*M*
W2 0.013 μ*M*
PbLuc 1.3 μ*M*
$HepG_2$ 24.1 μ*M*
KB 14.5 μ*M*

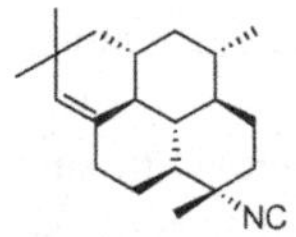

149
D6 0.25 μ*M*
W2 0.080 μ*M*
KB 48.8 μ*M*

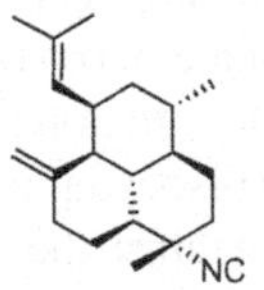

150 (isocyanoepiamphilectene)
D6 0.047 μ*M*
Dd2 0.99 μ*M*
W2 0.031 μ*M*
PbLuc 0.93 μ*M*
$HepG_2$ 28.0 μ*M*
KB 10.8 μ*M*

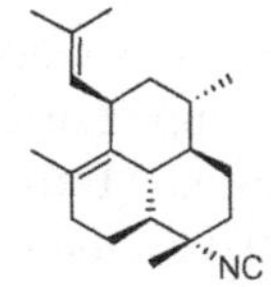

151
D6 0.20 μ*M*
W2 0.086 μ*M*
KB 51.2 μ*M*

The unusual β-lactam isonitriles monamphilectines B (**152**) and C (**153**) were isolated from the sponge *Svenzea flava*, and their structures and configurations were confirmed by semisynthesis from the known isonitrile 8,15-diisocyano-11(20)-amphilectene (**154**). The antiplasmodial activities of **152** and **153** were comparable to that of **154**, suggesting that the β-lactam group does not contribute much to this activity [361]. Kalihinol A (**155**) was first isolated from a sponge identified as an *Acanthella* sp. in 1984, but it was not until it was reisolated in 1998 that it was shown to have strong antiplasmodial activity with an IC_{50} of 1.2 n*M* against the FCR-3 strain of *P. falciparum* and a SI of 317 [362]. The related kalihinol B (**156**) was synthesized in 2015 and had comparable antiplasmodial activity [363]. A later paper from the same group reported the synthesis of simplified kalihinol analogs. Several of these derivatives retained strong antiplasmodial activity, including a di-isonitrile prepared in three steps from the economical precursor sclareolide. In addition, a preliminary stability study showed that at least one compound had good stability in the presence of liver microsomes, suggesting that isonitriles may have enough metabolic stability for use as antimalarial agents [364].

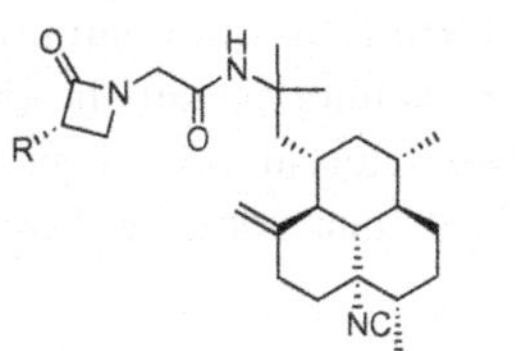

152 R = CH_3 (monamphilectine B)
3D7 44.5 n*M*
153 R = OCH_3 (monamphilectine C)
3D7 43 n*M*

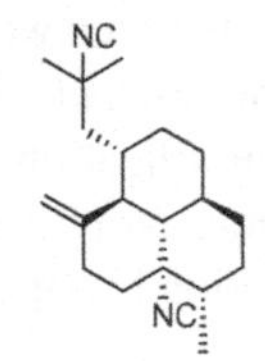

154 (8,15-diisocyano-11(20)-amphilectene)
W2 40 n*M*

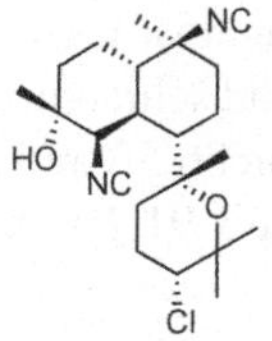

155 (kalihinol A)
FCR-3 1.2 n*M*
SI 317

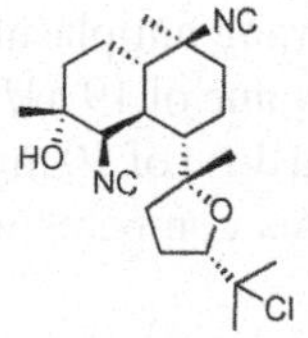

156 (kalihinol B)
3D7 8.4 n*M*,
Dd2 4.6 n*M*

The potent antimalarial activity of natural isonitriles prompted the investigation of the antimalarial activity of simple synthetic isonitriles, with the result that the isonitriles **157–159** were identified, with IC_{50} values of 88, 68, and 27 n*M* [365]. The SI of **159** was excellent, as it had been shown to be inactive against HaCat keratinocyte cells up to the maximum tested concentration of 128 μ*M* [366, 367], so these simple natural product-inspired compounds have potential as antimalarial agents.

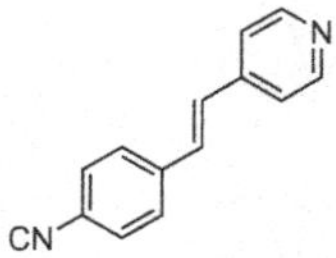

157
Dd2 88 n*M*

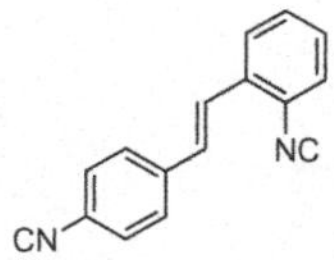

158
Dd2 68 n*M*

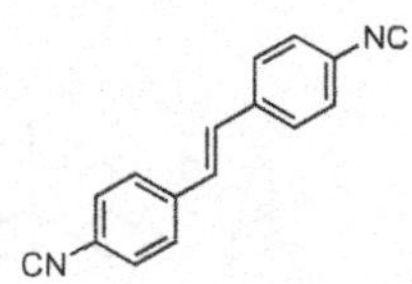

159
Dd2 27 n*M*
HaCat >128 μ*M*

Sesquiterpenoids

Other than artemisinin (**18**), discussed in Sect. 3.3, relatively few sesquiterpenoids have been identified as active antiplasmodial agents. Isomukaadial acetate (**160**) from *Warburgia salutaris* stem bark had an IC_{50} value of 1.4 μ*M* against the NF54 strain of *P. falciparum*, but the crude extract was much more potent at 0.01 μg/cm^3, indicating the presence of additional active compounds [368]. The spiro-sesquiterpenoid 3,4-dehydrotheaspirone (**161**) gave an IC_{50} of 27 n*M* against the 3D7 strain of *P. falciparum*, with a SI of >100 when compared with HL-60 cells, with its mechanism still under investigation [369]. The endoperoxide okundoperoxide (**162**) had only micromolar activity against several *P. falciparum* strains, so even though it had low cytotoxicity it is not a promising antiplasmodial agent [370].

Dobinin K (**163**) from the roots of *Dobinea delavayi* had good activity against *P. yoelli* BY265RFP in a 4-day in vivo assay, with 59% inhibition at 30 mg/kg/day, which compares favorably with artemisinin's 54% inhibition at 50 mg/kg/day. The depolarization of mitochondrial membrane potential with concomitant ROS production was proposed as a possible mechanism of antimalarial action [371]. The sesquiterpene lactone ineupatorolide A (**164**), isolated from *Carpesium rosulatum*, was found to have strong antiplasmodial activity against the D10 strain of *P. falciparum*, with an IC_{50} value of 19 n*M* [372]. It also had schizontocidal activity against *P. berghei* in mice at a dose of 10 mg/kg/day, with an 86.5% reduction in parasitemia in a four-day test. This compares with chloroquine's 91.2% reduction at a dose of 5 mg/kg/day [373].

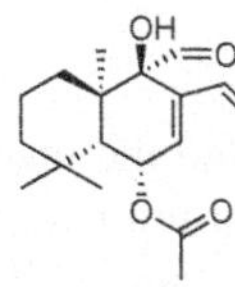

160 (isomukaadial acetate)
NF54 1.4 μ*M*

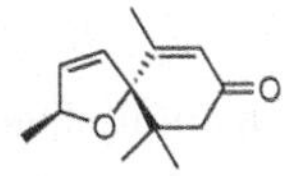

161
3D7 0.027 μ*M*
SI 3D7/HL60 100

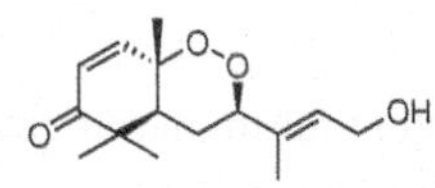

162 (okundoperoxide)
W2 1.75 μ*M*
D6 1.80 μ*M*
K1 5.59 μ*M*
NF54 4.88 μ*M*
cytotoxicity 85.0 μ*M*

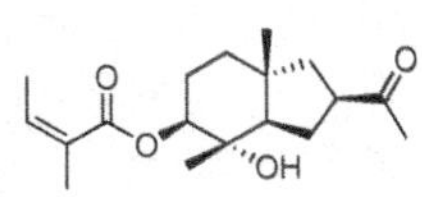

163 (dobinin K)

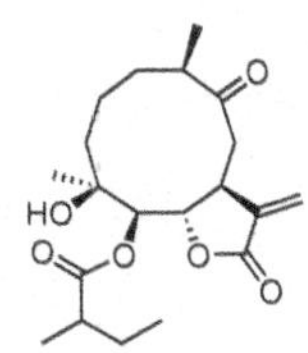

164 (ineupatorolide 1)
D10 0.019 μ*M*

Evaluation of a natural product assembly of lindenane-type sesquiterpenoids and sesquiterpenoid dimers for activity against the Dd2 strain of *P. falciparum* identified compounds fortunilide A (**165**), sarglabolide J (**166**), and chlorajaponilide C (**167**) as potent antiplasmodial agents, with IC_{50} values of 5.2, 7.2, and 1.1 n*M*, respectively, and SIs of 1700, 561, and 4900, when compared with the WI-38 (normal embryonic lung tissue) mammalian cell line. These compounds thus have excellent activity and selectivity, but their complex structures may limit their development [374].

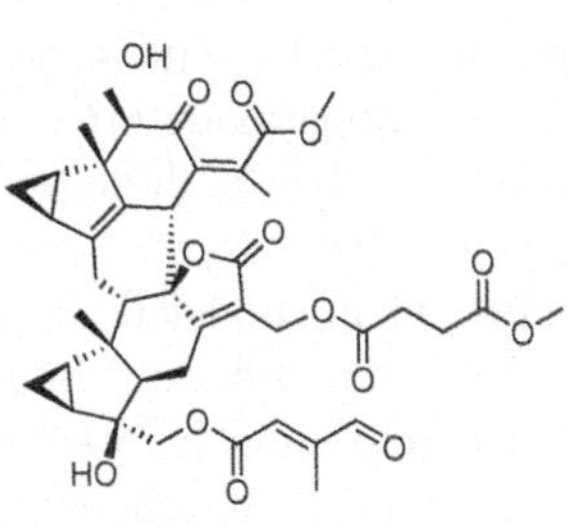

165 (fortunilide A)
Dd2 5.2 n*M*
WI-38 8.8 μ*M*
SI 1700

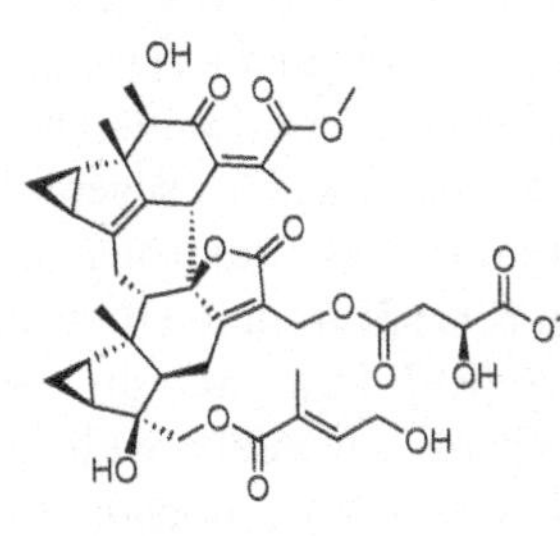

166 (sarglabolide J)
Dd2 7.2 n*M*
WI-38 4.04 μ*M*
SI 561

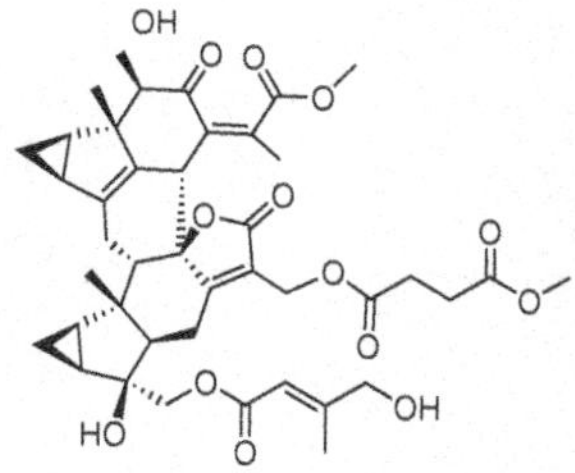

167 (chlorajaponilide C)
Dd2 1.1 n*M*
WI-38 5.39 μ*M*
SI 4,900

Diterpenoids

Several diterpenoids have shown promising antiplasmodial activity, but none has yet shown the right combination of potency, selectivity, and in vivo activity to be considered as a lead compound. As an example, the moderately antiplasmodial endoperoxide diterpene 13-*epi*-dioxiabiet-8(14)-en-18-ol (**168**) was isolated from *Hyptis suaveolens*, a plant used widely in traditional medicine. Its SI was not determined, so its major interest is its 70-fold greater activity than dehydroabietinol, a difference most probably due to its endoperoxide group [375]. However, the phenanthrenone derivative, fimbricalyx B (**169**), from the roots of *Strophioblachia fimbricalyx*, does have potent antiplasmodial activity with an IC_{50} of 19.3 n*M* against the K1 strain and an SI value of >500 when compared with Vero cells [376], and might thus be worth further investigation.

At first sight, ferruginol (**170**) does not appear to be a promising lead compound, with low micromolar activity against the K1 strain [377]. However, a SAR study of semisynthetic ferruginol derivatives identified a phthalimide (**171**) as the most potent compound, with IC_{50} values of 86 and 200 n*M* against the 3D7 and K1 strains and SIs of >25, so further structure–activity relationship studies as well as target identification studies were proposed [378].

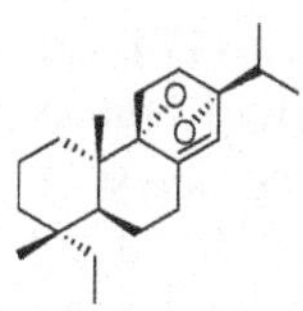

168 (13-*epi*-dioxiabiet-8(14)-en-18-ol)
D10 0.35 μ*M*

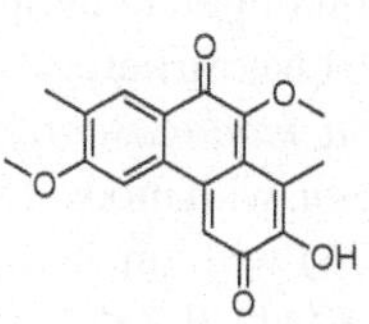

169 (fimbricalyx B)
K1 19.3 n*M*
Vero >10 μ*M*
SI >500

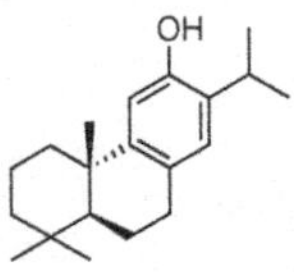

170 (ferruginol)
3D7 2.47 μ*M*
K1 1.33 μ*M*
HEK293 >25 μ*M*
SI >10

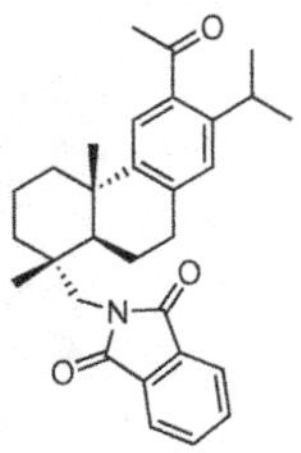

171
3D7 0.086 μ*M*
K1 0.201 μ*M*
HEK293 >25 μ*M*
SI >10

The diastereomeric aphadilactones C (**172**) and D (**173**) are the most potent of a series of related compounds isolated from the Meliaceae plant *Aphanamixis grandifolia*, with IC_{50} values of 170 and 120 n*M* against the Dd2 strain of *P. falciparum*. Aphadilactone C (**172**) also showed potent and selective inhibition against the diacylglycerol *O*-acyltransferase-1 (DGAT-1) enzyme [379]. The aphadilactones A–D were synthesized in 11 steps, including both inter- and intramolecular Diels–Alder cyclizations, in a 6.6% overall yield from known starting materials [380].

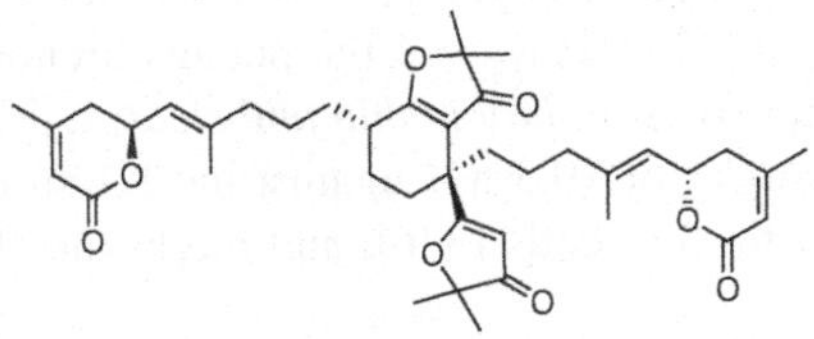

172 (aphadilactone C)
Dd2 0.17 μ*M*

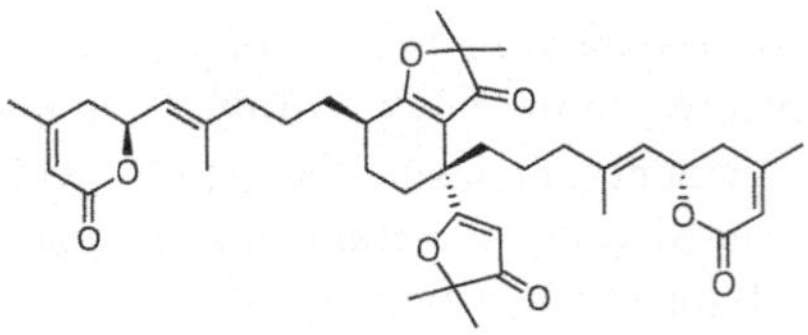

173 (aphadilactone D)
Dd2 0.12 μ*M*

Triterpenoids

As with the diterpenoids, relatively few triterpenoids have been reported to have the desired combination of potent antiplasmodial activity with low mammalian toxicity. As an example, the bisnortriterpenoids isoiguesterin (**174**) and 20-*epi*-isoguesterinol (**175**) from *Salacia madagascariensis* both showed high nanomolar IC_{50} values against *P. falciparum*, with modest SIs in a *P. falciparum*/Vero cell comparison [381]. Galtonoside C (**176**) with an IC_{50} value of 420 n*M*, not only had mid-nanomolar activity against the Dd2 *P. falciparum* strain but also had a CC_{50} value of 100 n*M* against the A2780 ovarian cancer cell line. Thus, it is not a potential lead compound [382].

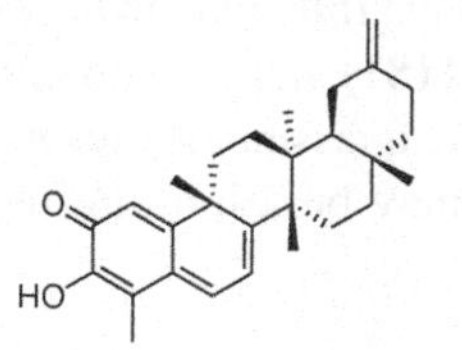

174 (isoiguesterin)
D6 495 n*M*
W2 420 n*M*
SI (D6) 3.4
SI (W2) 3.9

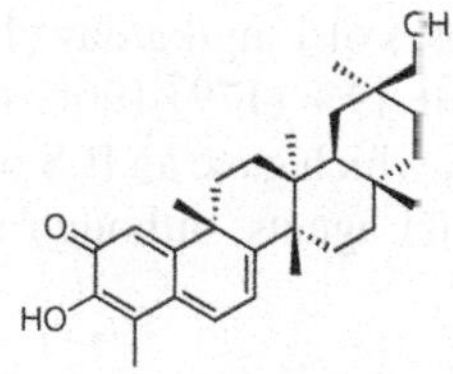

175 (20-*epi*-isoiguesterinol)
D6 56 n*M*
W2 56 n*M*
SI (D6 or W2) 31

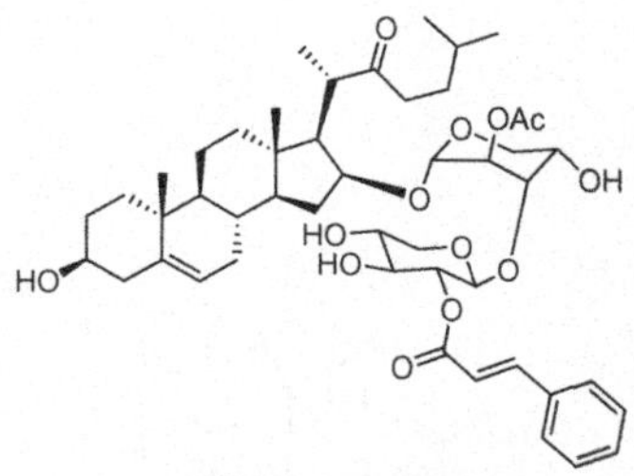

176 (galtonoside C)
Dd2 420 n*M*
A2780 100 n*M*
SI 0.24

The quassinoids, which are oxygenated degraded triterpene lactones, are a special case, and their antimalarial potential has been reviewed [383, 384]. Many quassinoids have potent antiplasmodial activity, but are also highly toxic. However, a few combine potent antiplasmodial activity with in vivo activity. One example is simalikalactone D (**177**), which has an IC_{50} of 10 n*M* against the FcB1 strain of *P. falciparum* and also inhibits 50% of *P. yoelii* rodent malaria parasites at an oral dose of 3.7 mg/kg/day [385]. Later studies showed that simalikalactone D had antiproliferative activities ranging from 6.3 n*M* for KB cells to 2 μ*M* on HeLa cells and 10 μ*M* on Vero cells, but that it was synergistic with atovaquone and may have potential as an adjuvant [386]. The known quassinoid eurycomanone (**178**) from *Eurycoma longifolia* was a very potent antiplasmodial agent, with an IC_{50} of 5 n*M* against the 3D7 strain, but no comparative cytotoxicity data are available, so it is not possible to assess its full potential as an antimalarial agent.

The modified quassinoids (**179–181**) were prepared by acylation of bruceolide in a SAR study of bruceolide derivatives [387]. The diacetyl derivative (**179**), which had been prepared previously [388], was almost completely metabolized in mouse serum over 180 min, and two carbonate derivatives (**180** and **181**) were prepared in an attempt to decrease this metabolism. The three derivatives had IC_{50} values of 39, 90, and 64 n*M*, respectively, against the FCR-3 strain of *P. falciparum*, and 16, >63, and 36 μ*M* against FM3A mouse mammary tumor cells, for respectable SIs of 410, >700, and 563. An in vivo 4-day suppressive test on mice infected with *P.*

berghei at doses of 1 mg/kg/day (**179**) or 3 mg/kg/day (**180**) and (**181**) gave increases in life span of 49% (**179**) 196% (**180**) and 182% (**181**), superior to chloroquine at 3 mg/kg/day, which gave an ILS of only 15%. These quassinoids thus have potential as antimalarial agents, although material supply may be a limiting factor in their development.

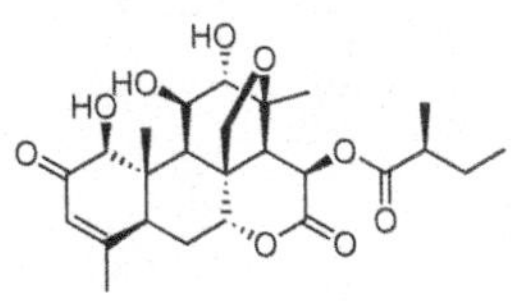

177 (simalikalactone D)
FcB1 10 n*M*
Vero 10 μ*M*
SI 1000

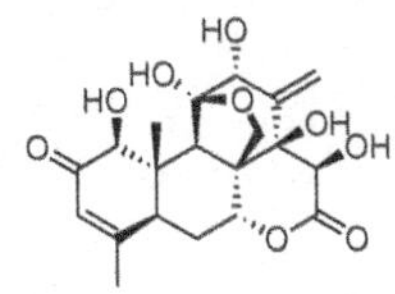

178 (eurycomanone)
3D7 5 n*M*

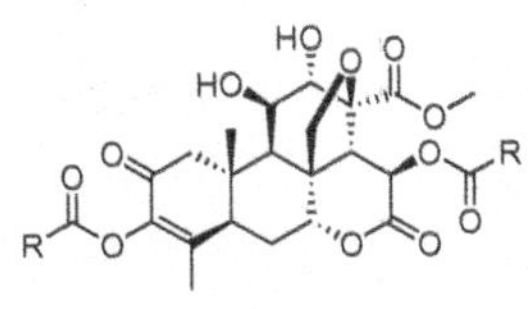

179 R = Me
FCR-3 39 n*M*
FM3A 16 μ*M*
SI 410

180 R = OMe
FCR-3 90 n*M*
FM3A >63 μ*M*
SI >700

181 R = OEt
FCR-3 64 n*M*
FM3A 36 μ*M*
SI 563

3.5.4 Shikimic Acid-Derived Metabolites

Many shikimic acid-derived metabolites with antiplasmodial activity have been isolated, but relatively few have significant levels of activity, and these are summarized here. A more complete listing can be found in the previously cited reviews, in particular that by Tajuddeen and van Heerden [245].

The stem bark of *Lophira lanceolata* yielded the cleaved flavonoid lophirone E (**182**) among other compounds. Although it only had weak antiplasmodial activity, with IC_{50} values of 38.5 and 12.2 μ*M* against the 3D7 and W2 strains, it had good gametocytocidal activity with an IC_{50} value of 0.14 μ*M* against stage V elo1CBG99 gametocytes. This stage-specific activity is unusual, and appears to be very specific, since the analog lophirone D (**183**) was a much less potent gametocytocidal agent. Lophirone E was only very weakly cytotoxic to the human dermal microvascular endothelial cell line HDMEC [389].

Licochalcone A (**184**) is an oxygenated chalcone isolated from the roots of the Chinese licorice plant [390]. It also had only weak antiplasmodial activity, but it was moderately effective in treating mice infected with *P. yoelli*, as it almost cleared parasitemia at the high dose of 50 mg/kg/day [391].

The culture broth of *Penicillium viticola* yielded several tropolones, with puberulic acid (**185**) showing the most potent activity with IC_{50} values of 51 and 50 n*M* against the K1 and FCR3 strains of *P. falciparum* and with SIs >5000 after evaluation with MCR-5 cells. The antiplasmodial activity appears to be dependent on having more than one free hydroxy group, since the corresponding methylated derivative (**186**) was about 80-fold less potent [392]. A recent study of new puberulic acid analogs identified the two esters **187** and **188**, among others, as promising lead compounds based on their inhibition of *P. bergei* in mice by the Peters' 4-day suppressive test at a dose of 15 mg/kg/day [393]. Somewhat surprisingly, these good in vivo results were obtained despite the poor SIs of these compounds.

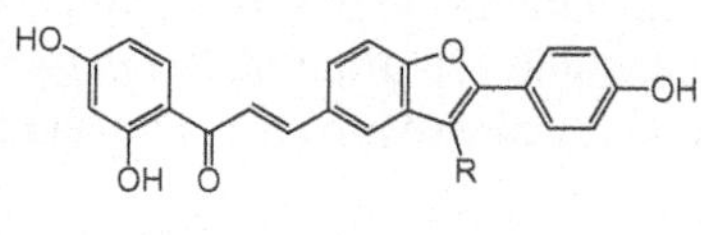

182 R = H (lophirone E)
3D7 38.5 μ*M*
W2 12.2 μ*M*
elo1CBG99 stage V gametocyte 0.14 μ*M*
HMEC-1 79 μ*M*
183 R = CHO (lophirone D)

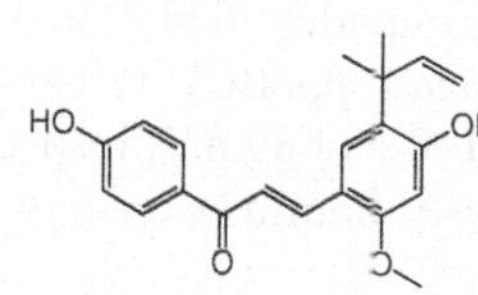

184 (lichochalcone A)
3D7 1.5 μ*M*
Dd2 ca.1.2 μ*M*

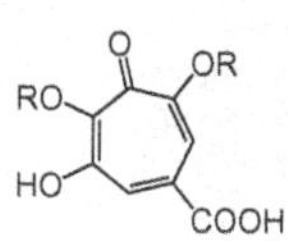

185 R = H (puberolic acid)
K1 51 n*M*
FCR3 50 n*M*
SI (MCR-5) >5000

186 R = Me
K1 3.95 μ*M*
SI (MCR-5) 8

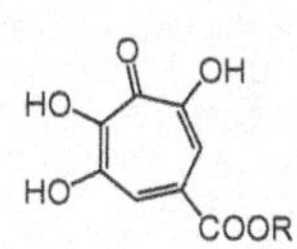

187 R = cyclohexyl
K1 85 n*M*
MRC-5 60 n*M*
SI 0.7

188 R = pentan-3-yl
K1 130 n*M*
MRC-5 208 n*M*
SI 1.6

The natural benzopyran encecalol angelate (**189**) was thought to be the antimalarial constituent of *Ageratum conyzoides*, but, although this turned out not to be the case [394], the synthetic chromene **190** based on the structure of **189** had good antiplasmodial activity [395, 396]. A recent investigation by the same group identified the chroman **191** and several related compounds as potent antiplasmodial agents, with an IC_{50} of 38 n*M* against the NF54 strain of *P. falciparum* and a SI of 171 for **191** compared with L6 rat skeletal myoblasts.

The presence of a phenolic hydroxy group appears to be necessary for activity, since compounds lacking this group were less potent antiplasmodial agents [397]. An in vivo evaluation of **191** in a 4-day test against *P. berghei* in mice failed to show significant reduction of the number of Plasmodia in mouse erythrocytes; this low vivo activity was explained by fast clearance from the blood plasma combined with the rapid biotransformation of **191** [397].

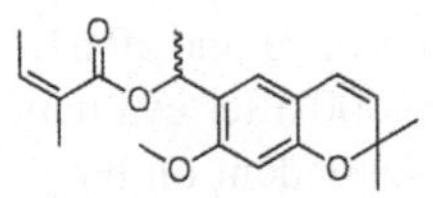

189 (encecalol angelate)

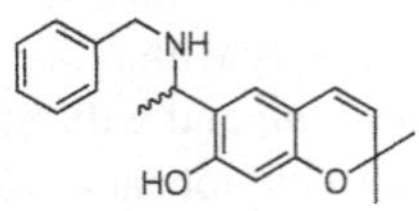

190
K1 20 n*M*
L6 109 μ*M*
SI 6760

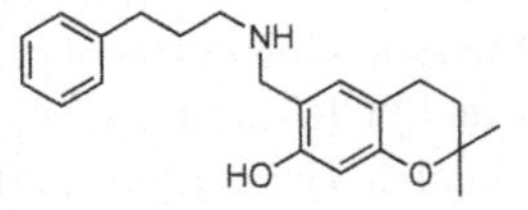

191
NF54 38 n*M*
SI (L6) 171

9-Methoxystrobilurin G (**192**) was the most potent antiplasmodial compound isolated from the basidiomycete *Favolaschia tonkinensis* BCC 18,689, with an IC_{50} value of 61 n*M* against the K1 strain of *P. falciparum* [398]. A later publication from the same group identified 9-methoxystrobilurins E (**193**), and K (**194**) from cultures of *Favolaschia* spp. BCC 18,686 as additional antiplasmodial compounds, with IC_{50} values of 140 and 89 n*M*, respectively, against the K1 strain of *P. falciparum.* The 9-methoxystrobilurins were only weakly active against Vero cells, and had SIs in the 1000 range [399].

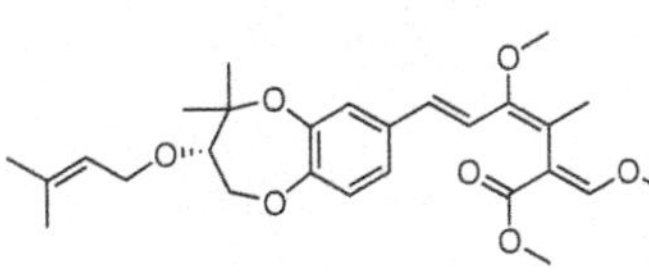

192 (9-methoxystrobilurin G)
K1 61 n*M*
Vero 50.7 μ*M*
SI 831

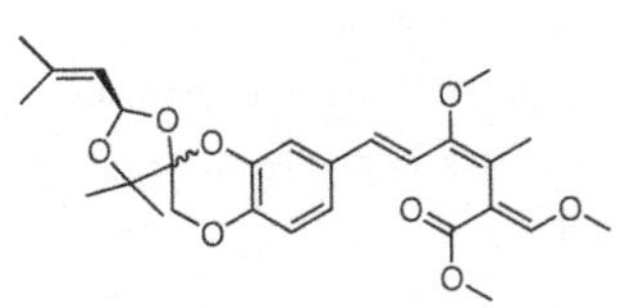

193 (9-methoxystrobilurin E)
K1 140 n*M*
Vero >103 μ*M*
SI >736

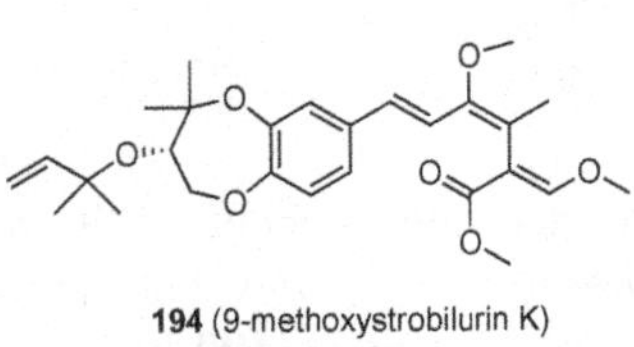

194 (9-methoxystrobilurin K)
K1 89 n*M*
Vero 98 μ*M*
SI 1100

The biflavanones GB-1 (**195**) and GB-2 (**196**) are the most potent compounds isolated from the seeds of the Congolese plant *Garcinia kola*, of which the nuts are used by traditional healers as a malaria preventative. GB-1 had an IC_{50} value of 0.16 μ*M* against *P. falciparum* FCR-3, a cycloguanil-resistant strain from Gambia, while the value for GB-2 was 0.21 μ*M*, and both compounds had SIs of >150, when compared with KB 3-1 cells. Oral treatment of *P. berghei*-infected mice with GB-1 in a 4-day test gave 39% inhibition of parasitemia at a dose of 25 mg/kg/day, compared with 53% inhibition by artemisinin at a dose of 15 mg/kg [400]. A study of kolaviron, which is a biflavonoid fraction from *Garcinia kola* seeds, showed that it had oral in vivo activity in *P. berghei*-infected mice [401].

Jacaglabrosides B–D (**197–199**) were isolated from the Ecuadorian plant *Jacaranda glabra*, a tree used in traditional folk medicine to treat leishmaniasis and other diseases. They had high nanomolar IC_{50} values against the K1 strain of *P. falciparum* and SIs of over 100, when compared with cytotoxicity to the L-6 rat skeletal myoblast cell line [402].

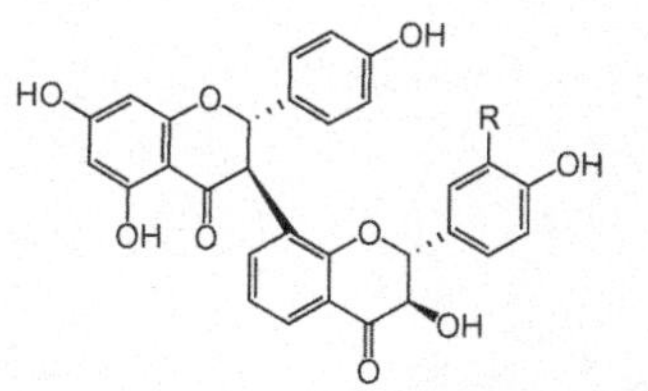

195 R = H (GB-1)
FCR-3 0.16 μ*M*
SI (KB 3-1) >900

196 R = OH (GB-2)
FCR-3 0.21 μ*M*
SI (KB3-1) > 700

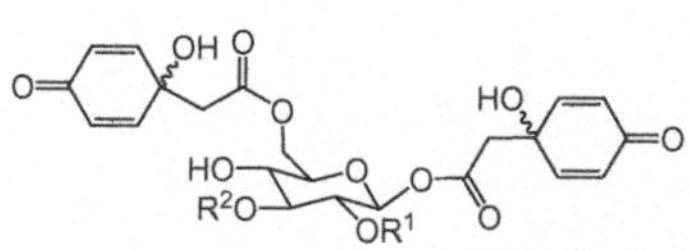

197 R^1 = B, R^2 = H (jacaglabroside B)
K1 0.93 μ*M*,
SI (L6) > 160 (L6)
198 R^1 = C, R^2 = B (jacaglabroside C)
K1 0.76 μ*M*
SI (L6) 155
199 R^1 = B, R^2 = C (jacaglabroside D)
K1 0.75 μ*M*
SI (L6) 154

B =

C =
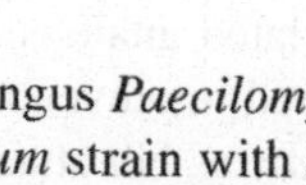

3.5.5 Aromatic Polyketides

Aigialomycin F (**200**) was reisolated from the fungus *Paecilomyces* sp., and found to be strongly potent against the 3D7 *P. falciparum* strain with an IC_{50} of 10.9 n*M* and also to have a CC_{50} value >50 μ*M* against Vero cells. Somewhat surprisingly it was essentially inactive against the Dd2 strain, so it has limited potential as a lead compound [403].

Racemic rhodomyrtosone F (**201**) was isolated as the major antiplasmodial constituent of *Syncarpia glomulifera* from Hawai'i, with an IC_{50} value of 100 n*M* against the Dd2 strain of *P. falciparum* and only 58% inhibition of HEK293 cells at a dose of 50 μ*M* [404]. It thus has modest potential as a lead compound, with an SI approaching 500.

Mallatojaponin C (**202**) was the most potent of three related compounds isolated from a Madagascar collection of *Mallotus oppositifolius*. It had an IC_{50} value of

140 n*M* against the Dd2 strain of *P. falciparum*, and it also had cytocidal activity against both the drug-sensitive HB3 strain and the drug-resistant Dd2 strain, with half maximal lethal dose (LD_{50}) values of 0.81 and 0.80 μ*M*. Since chloroquine has an LD_{50} value of 15.3 μ*M* against the Dd2 strain, this is an encouraging result. In addition, mallatojaponin C also had gametocytocidal activity, with an IC_{50} value of 3.6 μ*M* against gametocytes from the NF54 strain. However, these promising results were offset by the cytotoxicity of mallatojaponin C to A2780 human ovarian cancer cells, with a CC_{50} of 1.3 μ*M* [405]. Some simple analogs of mallatojaponin C were also prepared, and two of these had comparable antiplasmodial activities to the natural product [406].

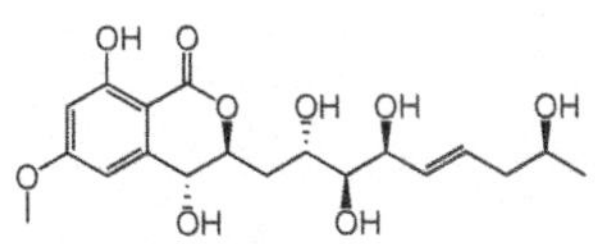

200 (aigialomycin F)
3D7 10.9 n*M*
Dd2 Not active
Vero >50 μ*M*
SI >4500

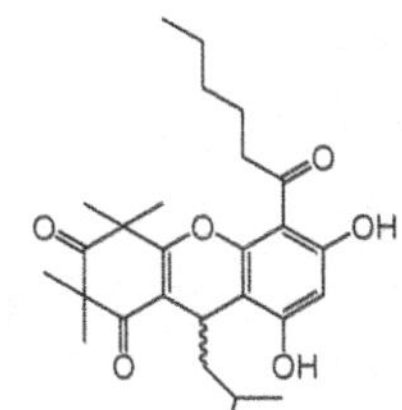

201 ((±)-rhodomyrtosone F)
Dd2 100 n*M*
SI (HEK293) ~400

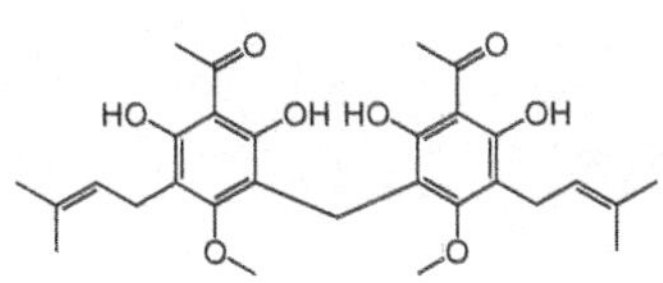

202 (mallatojaponin C)
Dd2 140 n*M*
A2780 1.3 μ*M*
SI 7.9

The caged *Garcinia* xanthone cluvenone (**203**) is a potent cytotoxic agent, but has only weak potency against *P. falciparum*. Synthesis studies led to the discovery that the A-ring hydroxylated analogs MAD28 (**204**) and MAD44 (**205**) had increased cytotoxicity compared to cluvenone [407]. MAD28 only had high sub-micromolar potency to *P. falciparum*, but attachment of an alkyltriphenylphosphonium bromide to the hydroxy groups gave compounds CR135 (**206**) and CR142 (**207**) with improved antiplasmodial activity and selectivity [408]. Compound **206** had IC_{50} values of 10.2 n*M* against the Dd2 strain and 12.3 n*M* against the 3D7 strain with SIs when compared with HEK293 cells of 142 and 118. The corresponding values for CR142 (**207**) were 15.0 and 18.1 n*M*, with SIs of 147 and 122 [408]. These values meet the minimum standard of potency and selectivity, but development will be hindered by the likely cost of the compound.

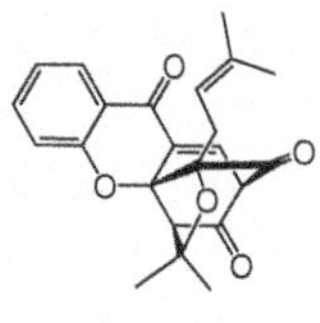

203 $R^1 = R^2 = H$
(cluvenone)
DD2 0.75 μ*M*

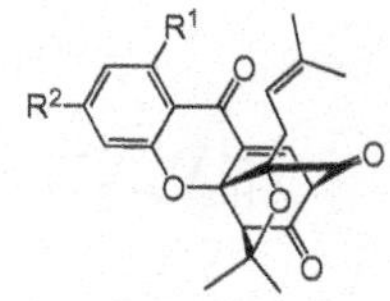

204 R^1 = OH, R^2 = H (MAD28)
Dd2 267 n*M*
SI (HEK293) 6.8
3D7 312 n*M*
SI (HEK293) 5.8

205 R_1 = H, R_2 = OH (MAD44)
Dd2 4.1 μ*M*

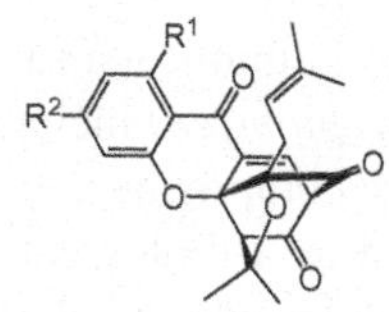

206 $R^1 = O(CH_2)_4P^+Ph_3Br^-$, R^2 = H (CR135)
Dd2 10.2 n*M*
SI (HEK293) 142
3D7 12.3 n*M*
SI (HEK293) 118

207 R^1 = H $R^2 = O(CH_2)_4P^+Ph_3Br^-$ (CR142)
Dd2 15.0 n*M*
SI (HEK293) 147
3D7 18.1 n*M*
SI (HEK293) 122

Various bisanthraquinones have shown antiplasmodial activity, but in many cases, such as for chryslandicin (**208**) and its congener **209**, the activity is not significantly selective [409]. The joziknipholones A (**210**) and B (**211**), however, combine moderate antiplasmodial activity to the K1 strain of *P. falciparum*, with IC_{50} values of 170 and 260 n*M*, and SIs of 96 and 67 [410]. The unusual sulfur-containing quinones urdamycinone E (**212**) and G (**213**) were isolated from a *Streptomyces* sp. collected in Thailand, and have submicromolar activity combined with modest SIs [411]. Synthesis studies to develop accessible antiplasmodial anthraquinones met with only partial success, with the caffeic acid derivative of aloe emodin having an IC_{50} value of 1.3 μ*M* [409].

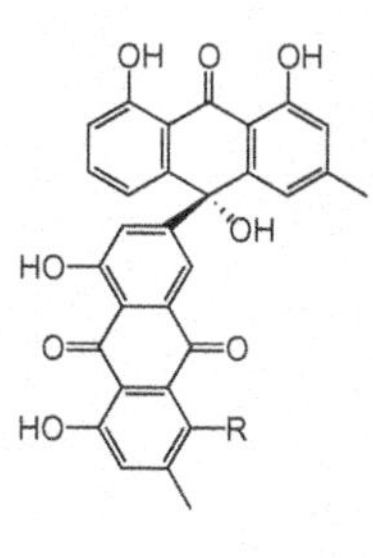

208 R = OH (chryslandicin)
Dd2 0.2 μ*M*
A2780 4 μ*M*
SI 20

209 R = H
Dd2 0.4 μ*M*
A2780 6 μ*M*
SI 15

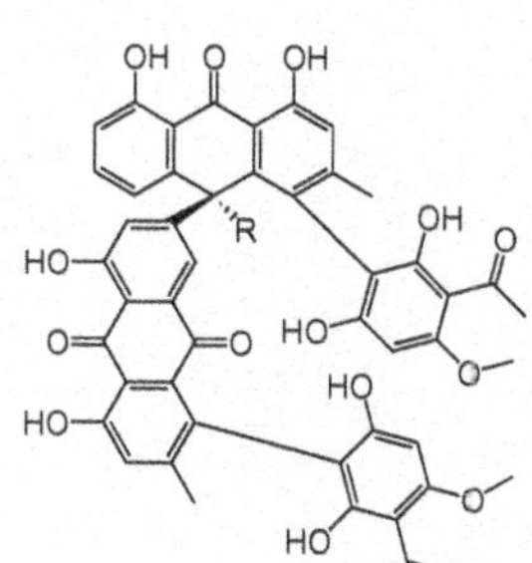

210 R = H (joziknipholone A)
K1 170 n*M*
L6 16.3 μ*M*
SI 96

211 R = OH (joziknipholone B)
K1 260 n*M*
L6 17.4 μ*M*
SI 67

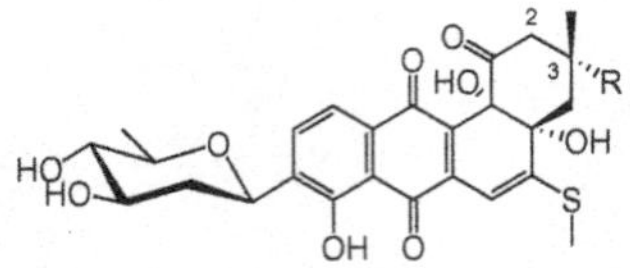

212 R = OH (urdamycinone E)
K1 99 n*M*
SI 32

213 (urdamycinone G $\Delta 2,3$)
K1 270 n*M*
SI 21

3.5.6 Macrolides

Bafilomycin A_1 (**214**) is a macrolide antibiotic with a 16-membered ring, and has been isolated from various species of *Streptomyces* including *Streptomyces spectabilis.* It has a variety of biological activities, comprising antitumor, immunosuppressant, and antifungal effects, and it was used as a biological tool compound to confirm the presence of V-ATPase in food vacuoles and the plasma membrane of *P. falciparum* cells [412, 413]. This inhibition is the most likely explanation of its ability to inhibit *P. falciparum* with an IC_{50} value of 66 n*M* against the K1 strain [343]. However, its CC_{50} of 1.72 μ*M* against Vero cells indicates that it is an unlikely candidate for development as an antimalarial agent.

Bastimolide A (**215**) is a 40-membered ring macrolide that was isolated from the tropical marine cyanobacterium *Okeania hirsuta* [414]. It showed good antiplasmodial activity against four resistant strains of *P. falciparum,* with IC_{50} values of 0.08, 0.09, 0.14 μ*M*, and 0.27 μ*M* against the TM90-C2A, TM90-2CB, W2, and TM91-C235 strains while having a weaker activity against Vero cells with a CC_{50} value of 2.1 μ*M* [414].

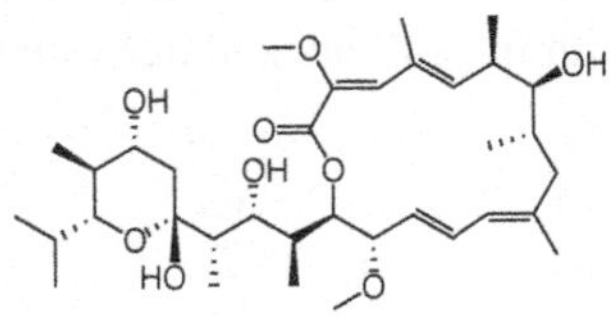

214 (bafilomycin A_1)
K1 66 n*M*
Vero 1.72 μ*M*
SI 26

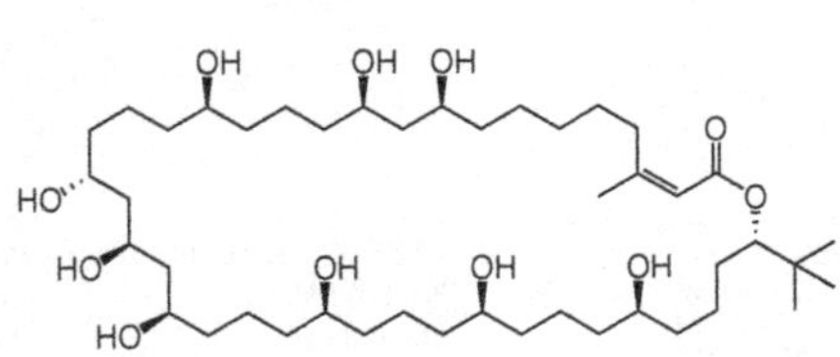

215 (bastimolide A)
TM90-C2A 80 n*M*
W2 140 n*M*
Vero 2.1 μ*M*
SI (W2) 15

Concanamycin A (**216**) was isolated from the mycelium of *Streptomyces diastatochromogenes S*-45 as an 18-membered macrolide lactone with potent antibacterial activity [415]. Like bafilomycin A_1, it is an inhibitor of vacuolar H^+-ATPase, and also has shown fungicidal, larvicidal, and cytotoxic activities. It had an IC_{50} value of 0.2 n*M* against the K1 strain of *P. falciparum*, with an additive effect in combination with pyronaridine [416]. However, its subnanomolar activity against UMR-106 rat osteoblasts [417] suggests that has a low selectivity for plasmodial cells, and will thus be difficult to develop as an antimalarial agent.

Romidepsin (**217**) was isolated from *Chromobacterium violaceum* No. 968 as FR901228 [418]. It had nanomolar level cytotoxic activity against several cancer cell lines and to HE-9 human endothelial cells, but was not strongly cytotoxic to normal human fibroblast HNF cells [418]. It was then developed as an anticancer drug, and is approved for use in Europe and the USA for the treatment of T-cell lymphomas. A 2015 study of it and other HDAC inhibitors for antiplasmodial activity indicated that it had an IC_{50} value of 90 n*M* against *P. falciparum* strain 3D7 and 130 n*M* against the Dd2 strain [419], and a study of *Chromobacterium* spp. identified romidepsin as the major antiplasmodial constituent [420]. However, since it also had low nanomolar cytotoxicity to NFF and HEK 293 mammalian cells, its potential as an antimalarial agent is unclear.

The endophytic fungus *Strasseria geniculata* CF-247251 produced the four macrolides strasseriolides A–D, with strasseriolide B (**218**) having the most potent activity against *P. falciparum* 3D7 in a lactate dehydrogenase assay, with an IC_{50} value of 13 n*M* and no significant cytotoxicity against the HepG2 cell line. It was less potent in the standard SYBR Green assay, with an IC_{50} value of 162 n*M* against the 3D7 strain, but it had an IC_{50} value of 32 n*M* against the Dd2 strain. This, coupled with its good selectivity, makes it a potential lead compound [421].

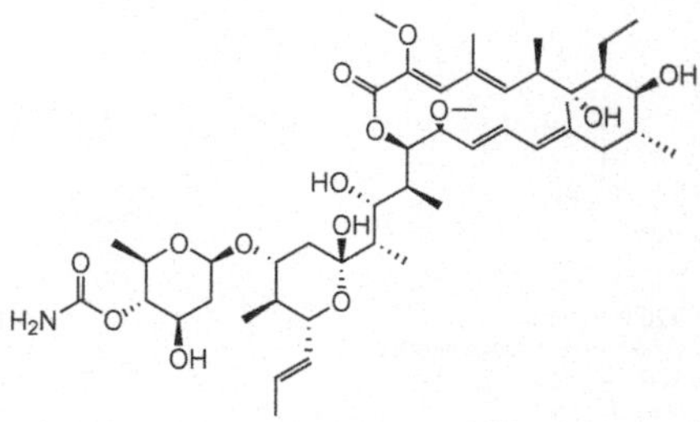

216 (concanamycin A)
K1 0.2 n*M*
UMR-106 0.5 n*M*
SI 2.5

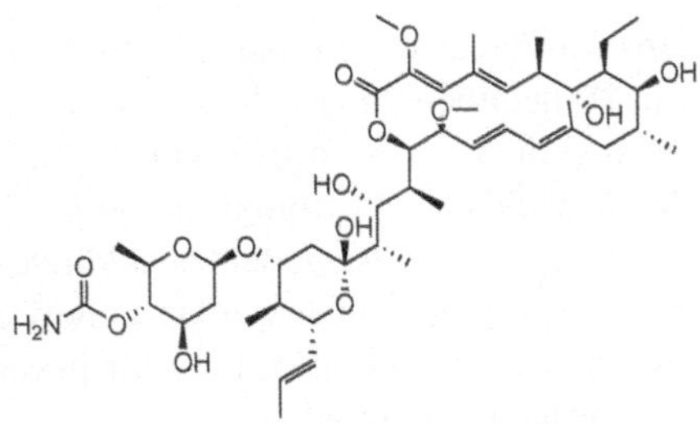

216 (concanamycin A)
K1 0.2 n*M*
UMR-106 0.5 n*M*
SI 2.5

Bromophycolide A (**219**) is an unusual brominated macrolide from the Fijian red macroalga *Callophycus serratus*, and was found to have good antiplasmodial activity against the 3D7 strain of *P. falciparum*, with an IC_{50} value of 499 n*M* and an SI of 69 compared with Vero cells [422]. Both potency and selectivity were improved in the acetate derivative **220**, and this finding was then used to append two immunoaffinity fluorescent tags. Studies with tagged **219** then showed that it inhibited heme crystallization as its primary mechanism of action [423].

Aigilomycin (**221**) and the related paecilomycins E (**222**) and F (**223**) were isolated from a fungus of the *Paecilomyces* genus [403]. Aigilomycin and paecilomycin E both had submicromolar activity against the *P. falciparum* 3D7 strain, but were much less potent towards the Dd2 strain, while paecilomycin F only had micromolar potency to both strains. Both of the paecilomycins were relatively non-toxic to Vero cells, with CC_{50} values above 50 μ*M*. Their weak activity against the drug-resistant Dd2 strain makes them unlikely lead compounds, however.

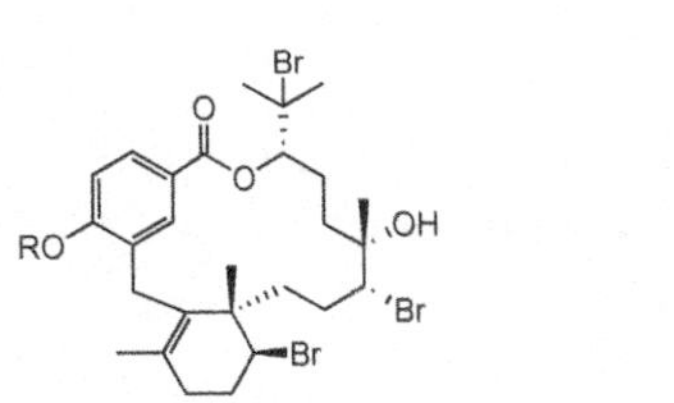

219 R = H
(bromophycolide A)
3D7 499 n*M*
Vero 34.6 μ*M*
SI 69

220 R = OAc
(bromophycolide A acetate)
3D7 241 n*M*
Vero 74 μ*M*
SI 307

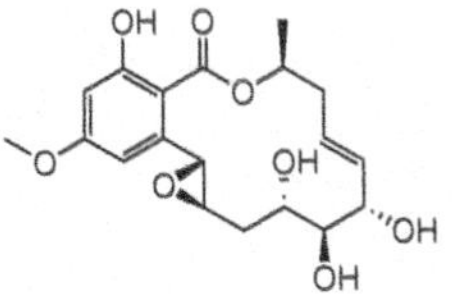

221 (aigilomycin B)
3D7 650 n*M*
Dd2 13.5 μ*M*

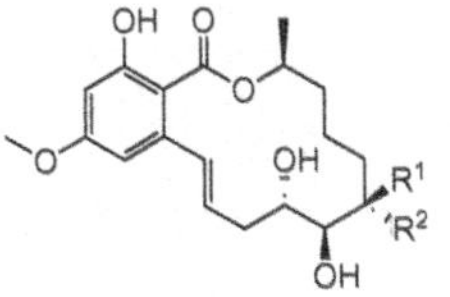

222 R^1 = OH, R^2 = H
(paecilomycin E)
3D7 20 n*M*
Dd2 8.8 μ*M*
Vero > 50 μ*M*

223 R^1 = H, R^2 = OH
(paecilomycin F)
3D7 1.1 μ*M*
Dd2 1.7 μ*M*
Vero > 50 μ*M*

3.5.7 Peptides

Peptides, both cyclic and linear, have proved a fruitful source of antiplasmodial compounds, and information on the more promising ones for development is summarized below.

Actinoramide A (padanamide A) (**224**) was isolated from a marine *Streptomyces* sp. in 2011 [424, 425] and identified as an antimalarial lead from a collection of natural product extracts using a quantitative high-throughput screening (qHTS) paradigm [426]. It had IC_{50} values in the 0.19–0.34 μM range against five different *P. falciparum* lines, and only low activity against HEK293 cells, with an SI of at least 500 [426]. Its modest antiplasmodial activity makes it a less than ideal development candidate, but its good SI suggests that there is potential for structural optimization.

Aerucyclamide B (**225**) is a modified hexacyclopeptide that was isolated along with its congener, aerucyclamide A, from the toxic freshwater cyanobacterium *Microcystis aeruginosa* PCC 7806 [427]. Both compounds were toxic to the freshwater crustacean *Thamnocephalus platyurus*, with LC_{50} values of 30.5 and 33.8 μM. It has an IC_{50} value of 0.7 μM against *P. falciparum* K1, and 120 μM against rat myoblast L6 cells, so it is reasonably selective for plasmodial cells although its antiplasmodial activity is only moderate [428].

Cyclomarin A (**226**) was discovered by a large cell-based screening program of over a million synthetic compounds and natural products for growth inhibition of *P. falciparum* blood stage, which identified over 6000 compounds with good activity [429]. One of the compounds was the cyclic peptide cyclomarin A, which was isolated along with cyclomarins B and C from the streptomycete strain CNB-982 from a marine sediment sample collected in Mission Bay, CA. It is only moderately cytotoxic, with a mean CC_{50} value of 2.6 μM against a panel of human cancer cell lines [430], but it inhibits the growth of blood stage *P. falciparum* (NF4 strain) with an IC_{50} value of 40 nM. Its molecular target, diadenosine triphosphate hydrolase (PfAp3Aase), was identified by chemical proteomics. It had an IC_{50} value of 4 nM against this enzyme, but >10 μM against the related enzyme PfAp4Aase and the closest human homologue hFHIT, so it is likely to be selective for *P. falciparum*, and has been identified as a starting point for drug discovery [431].

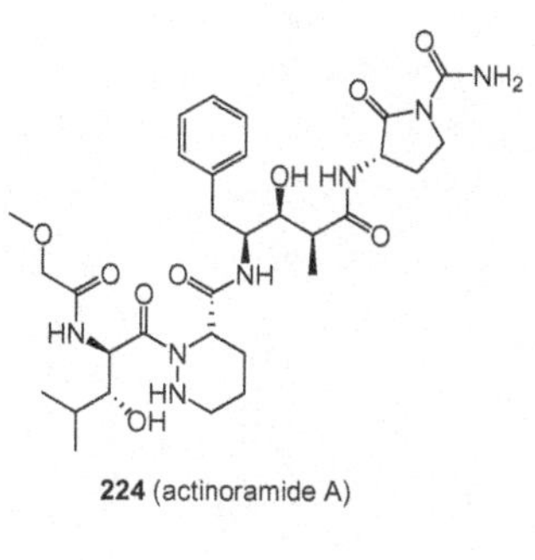

224 (actinoramide A)

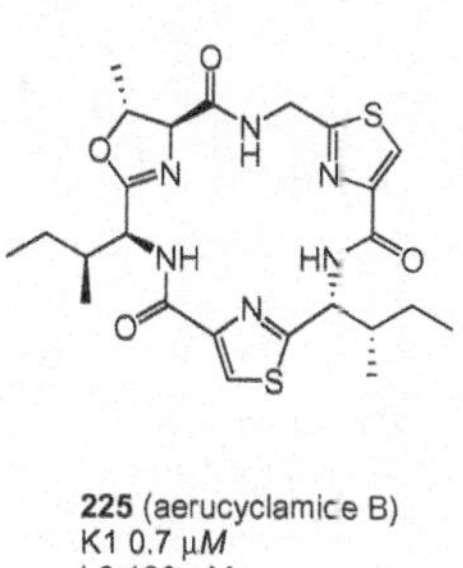

225 (aerucyclamice B)
K1 0.7 μM
L6 120 μM
SI 171

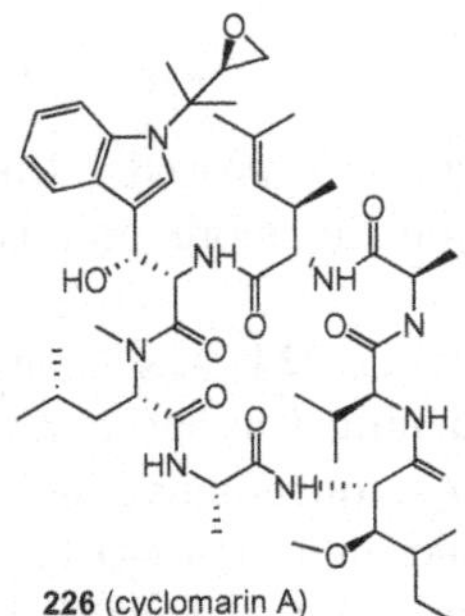

226 (cyclomarin A)
NF54 40 n*M*
human cell panel 2.6 μ*M*

Echinomycin (**227**) was first isolated from *Streptomyces echinatus* in 1959 [432], and was shown to be a DNA-binding compound that intercalates DNA by a bifunctional mode [433]. It inhibits the growth of *P. falciparum* with an IC_{50} value of 3.7 n*M*, while having a CC_{50} of 4.4 μ*M* against human mammary epithelial cells [434]. It is thus a potent and selective antimalarial agent, and could be a potential base molecule for synthetic medication to improve antimalarial activity and to reduce mammalian cell toxicity. The related compound kakadumycin A has similar activities [434], with an IC_{50} value of 6.4 n*M* against *P. falciparum* and a CC_{50} of 7.3 μ*M*, but its structure has not yet been published in the scientific literature.

Tyrocidine A (**228**) is a cyclic peptide that is the most potent antimalarial component of the tyrothricin complex, isolated from *Bacillus brevis.* The tyrothricin complex was shown to have antiplasmodial activity as early as 1944 [435], and a 2007 study reported the isolation and antiplasmodial evaluation of **228** [436]. It had an IC_{50} value of 0.58 n*M* against the 3D7 strain of *P. falciparum*, with CC_{50} values of 14 μ*M* against HeLa and 21 μ*M* against A549 mammalian cells, so it has very good selectivity for the parasite. Its activity was not due to lysis of infected erythrocytes, but to inhibition of parasite development and life cycle progression.

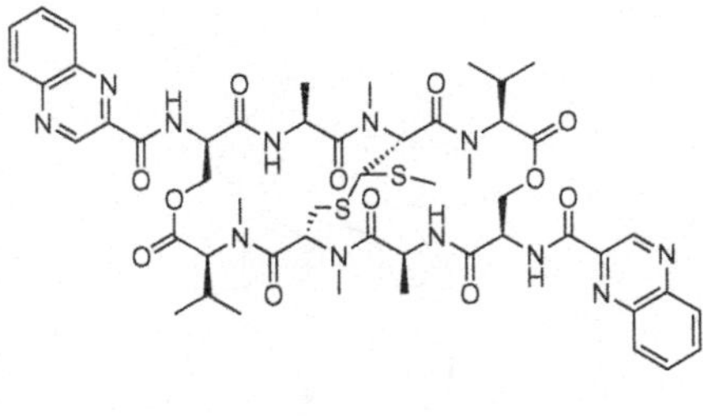

227 (echinomycin)
CSC-1 3.7 n*M*
4.4 μ*M*

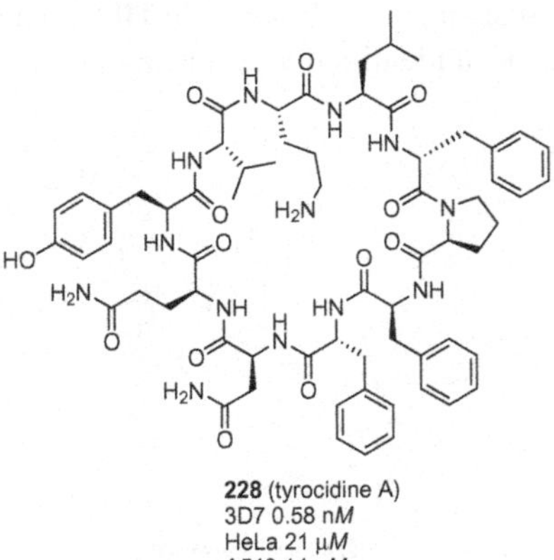

228 (tyrocidine A)
3D7 0.58 n*M*
HeLa 21 μ*M*
A549 14 μ*M*

Gramicidin D is a mixture of the linear peptides gramicidins A, B, and C, with gramicidin A (**229**) comprising about 80% of the total. Each of the individual peptides exists in two isoforms, and has alternating L- and D-amino acids. It was discovered in 1930, and played a role in stimulating research on penicillin [437]; it is still used for topical application and in eye drops, usually in combination with other antibiotics. It exists in a helical form, and thus acts as an ion-channel quasi-ionophore. It has an exquisitely potent IC_{50} of 0.018 n*M* against the Nigerian strain of *P. falciparum*, with CC_{50} values of 15 and 35 n*M* against U937 leukemia and Jurkat cells, respectively, for SIs of 830 and 1950 [438]. It has not been developed as an antimalarial agent, however, most probably because of its generalized toxicity due to hemolysis [439], but it had in vivo activity against *P. vinckei petteri* in mice with an ED_{50} of 1.4 mg/kg and a therapeutic index of 344 when administered intraperitoneally; unfortunately it was not orally active [440]. It was one of several natural products identified in a chemical genomic profiling approach to antimalarial drug discovery, with an average IC_{50} value of 1.52 n*M* against multiple *P. falciparum* isolates [441].

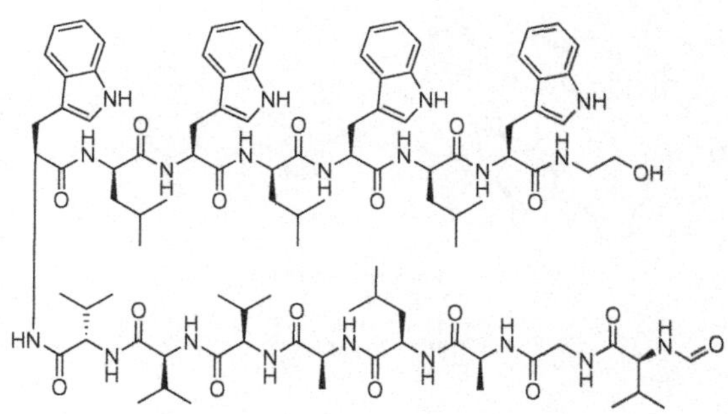

229 (gramicidin A)
Pf 0.018 n*M*
U937 15 n*M*
Jurkat 35 n*M*

The cyclic peptide lyngbyabellin A (**230**) was found as the most potent antiplasmodial component of the Fijian marine cyanobacterium *Moorea producens* [442]. It had an IC_{50} value of 0.0015 μ*M* against blood-stage *P. falciparum* and a CC_{50} value of 19 and 3.3 μ*M* against HEK293T and HepG2 cells, respectively, so it has reasonable selectivity. It was not, however, strongly active against liver-stage *P. berghei*. Other cyclic peptide or macrolide congeners were less potent or less selective for *P. falciparum*.

Micrococcin P1 (**231**) is an unusual pyridinyl polythiazole peptide that was first reported in 1948 as micrococcin from a *Micrococcus* sp. Its structure elucidation was complicated by challenging stereochemical issues, and its structure was only assigned conclusively as shown below after a series of total syntheses, as reviewed by Ciufolini and LeFranc [443]. Micrococcin P1 (**231**) has an IC_{50} value of 35 n*M* against *P. falciparum* 3D7 [444], and only weak cytotoxicity against the hepatic cell line HepG2 and the monocytic cell line THP-1, for a SI of at least 850 [445]. It is thus a potential platform for the synthesis of analogs with improved antimalarial activity and selectivity.

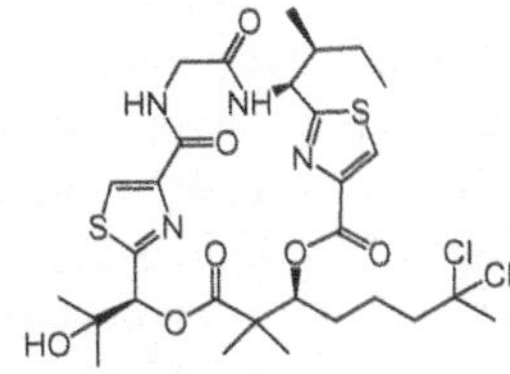

230 (lyngbyabellin A)
Dd2L 1.5 n*M*
HEK293T 19 μ*M*
HepG2 3.3 μ*M*
SI (Dd2L) 12,500
SI (HEK293T) 2200

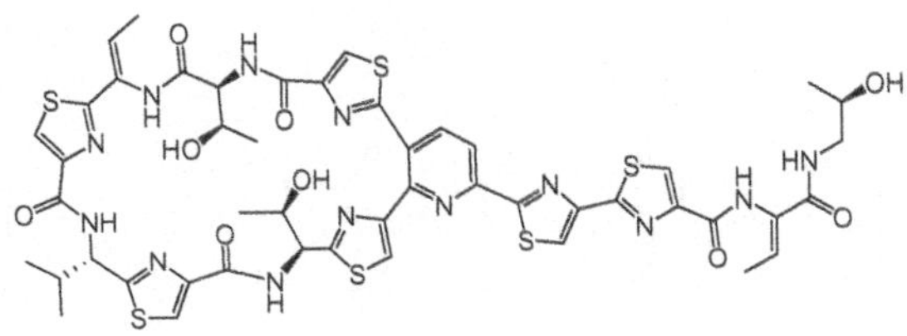

231 (micrococcin P1)
3D7 35 n*M*
HepG2 >30 μ*M*
THP-1 >30 μ*M*
SI >850

3.5.8 Polyether Ionophores

Twenty-two polyether ionophore compounds were screened for their antiplasmodial activities against the Nigerian strain of *P. falciparum* by Gumila et al. in 1996, and lasolacid A (**232**), lonomycin (**233**), monensin (**234**), narasin A (**235**), salinomycin (**236**), and nigericin (**237**) each stood out as having excellent potency and selectivity [438]. The relevant data are shown in Table 2 and indicate that these compounds have reasonable potency and moderate selectivity. A later paper from the same group showed that mice infected with *Plasmodium vinckei petteri* or *Plasmodium chabaudi* could be treated intraperitoneally with these ionophores, with 5-bromolasolacid A having a therapeutic index of 18 and remaining active after oral administration [440]. A later study evaluated salinomycin (**236**), monensin (**234**), and nigericin (**237**) for activity against the D10, W2, 3D7, Ghana, and Burkina Faso strains of asexual *P. falciparum* parasites, with both **234** and **237** showing low nanomolar potencies. The authors concluded that "our data suggest a strong and fast activity of monovalent ionophores against asexual stages and against mature gametocytes and ookinetes, resulting in a block of sporogonic development in the mosquito. These compounds, being active against several different *P. falciparum* life stages, thus appear to be

Table 2 Antiproliferative and cytotoxic activities of selected polyether ionophores[a]

Compound	IC_{50}/nM for *P. falciparum*	MV_{50}/nM for macro-phage U937 cells	LV_{50}/nM for Jurkat lymphoblast	In vitro differential for U937 cells	In vitro differential for Jurkat cells
232 Lasolacid A	49	1740	2430	35.7	50
233 Lonomycin	1.7	284	398	168	236
234 Monensin A	2.2	95	260	43	117
235 Narasin A	1.3	400	650	307	500
236 Salinomycin	3.6	–	–	–	–
237 Nigericin	1.4	260	120	180	90

[a]MV_{50} and LV_{50} are the drug concentrations leading to 50% inhibition of ^{3}[H]thymidine incorporation into the DNA of macrophages or lymphoblasts, respectively

particularly attractive as potential curative and transmission-blocking agents" [446]. Narasin A (**235**) was among a number of compounds identified as antimalarial by chemical genomic profiling, where it had a mean IC_{50} value of 0.112 μM against a panel of *P. falciparum* isolates [441]. The combination of nigericin (**237**) with monensin (**234**) was shown to be synergistic for the treatment of mice infected with *P. vinckei petteri* [447].

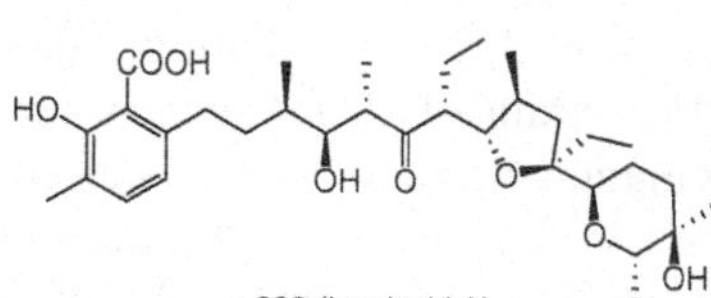

232 (lasolacid A)

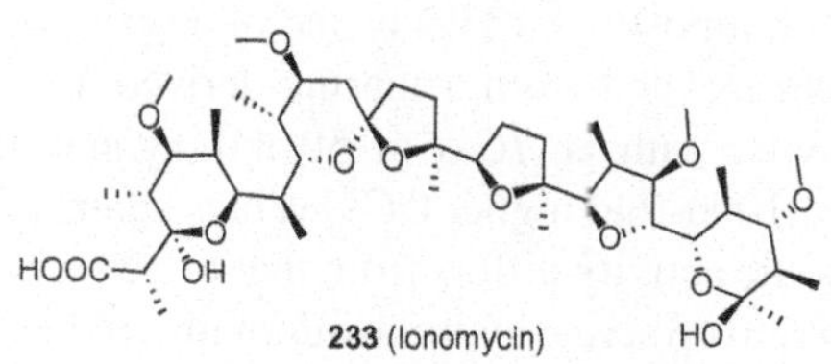

233 (lonomycin)

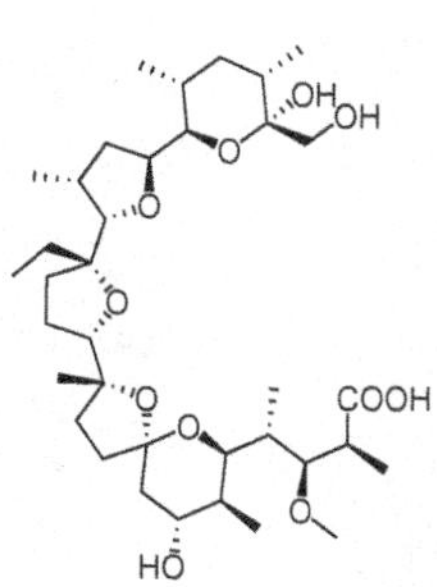

234 (monensin)

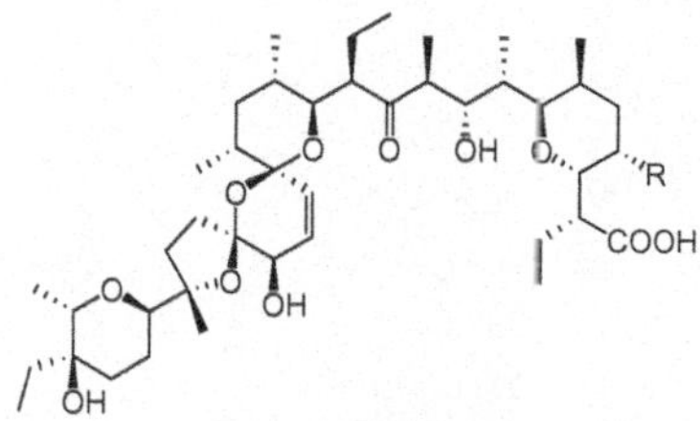

235 R = Me (narasin A)
236 R = H (salinomycin)

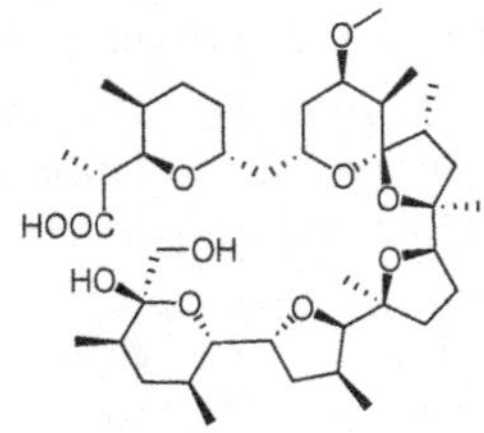

237 (nigericin)

3.5.9 Miscellaneous Compounds

Cladosporin (asperentin) (**238**) is a fungal isocoumarin first isolated from *Cladosporium cladosporioides* in 1971, where it was identified as an antifungal agent [448]. It was later identified as a potent antiplasmodial agent with activity against both blood- and liver-stage diseases with IC_{50} values in the 40–90 nM range, and a CC_{50} values in the 10–70 μM range for three mammalian cancer cell lines. It was also shown to be a potent and specific inhibitor of cytosolic lysyl-tRNA synthetase in *Plasmodium* spp. [449]. It is >100-fold more potent against parasite lysyl-tRNA synthetase relative to the human enzyme, and it is thus a potential lead compound with a novel antimalarial target. Its poor metabolic stability precludes its direct use as an antimalarial agent, but several analogs have been investigated to understand its SAR and potentially develop improved analogs with clinical prospects [450].

Salinipostin A (**239**) is one of a series of marine long-chain bicyclic phosphotriesters isolated from a marine-derived *Salinospora* sp. It has good antiplasmodial activity, with an IC_{50} of 50 nM against the W2 strain of *P. falciparum* and at least 1000-fold higher CC_{50} values against four mammalian cell lines, and its stage-specific activity differs from that of heme binders [451]. Its targets in the parasite all contain α/β serine hydrolase domains, indicating that these are promising antimalarial targets [452].

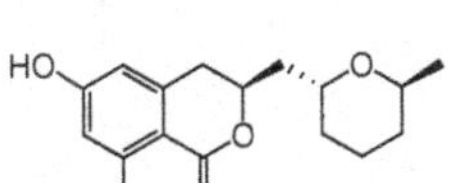

238 (cladosporin)
3D7 45 nM
K1 80 nM
HepG2 43.6 μM
SI (K1) 545

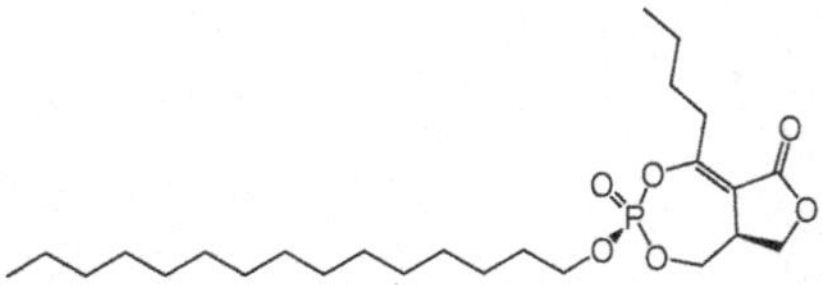

239 (salinipostin A)
W2 0.05 μM
HEK293T >50 μM
SI >1000

Methionine aminopeptidase 2 (MetAP2) is a key housekeeping enzyme in the biosynthesis of many proteins, and its gene has been cloned from the malaria parasite *P. falciparum* [453]. The microbial product fumagillin (**240**) and its semisynthetic analog TNP-470 (**241**) have good activities against *P. falciparum*, with IC_{50} values in the low nanomolar range for fumagillin and low to subnanomolar values for TNP-470 [453]. Both fumagillin (**240**) and TNP-470 (**241**) were shown to bind covalently to MetAP2, while the new analog fumarranol **242** binds non-covalently. Both **242** and **241** were tested in mice infected with the lethal CQ-resistant *P. yoelii* 17X strain in a 4-day test. Although **242** was less effective than TNP-470 at 20 mg/kg/day or than chloroquine (**14**) at 10 mg/kg/day, it was curative at 120 mg/kg/day in three out of five mice, so it is likely that further analogs could yield improved treatments [454]. The role of metalloaminopeptidases as targets for antimalarial drugs was discussed in a recent review [455].

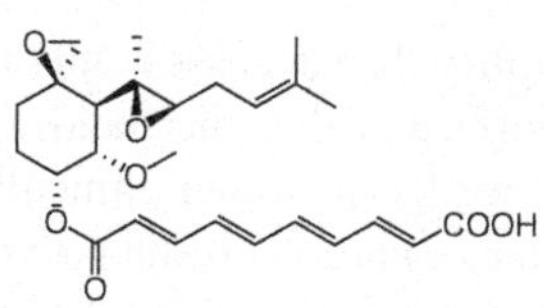

240 (fumagillin)
D6 38 *nM*
W2 17 *nM*
3D7 9 *nM*

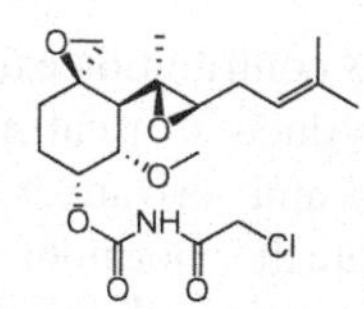

241 (TNP-470)
D6 4.1 *nM*
W2 3.7 *nM*
3D7 0.75 *nM*

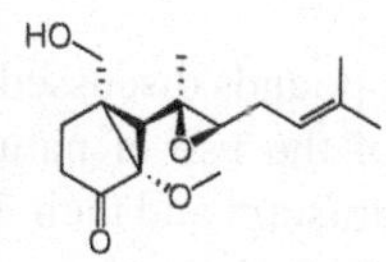

242 (fumarranol)
NF54 0.16 μ*M*
W2 0.27 μ*M*

Plakortin (**243**) and epiplakinidoic acid (**244**) are the most potent of several endoperoxides isolated from the marine sponges *Plakortis simplex* [456] and *P. halichondrioides* [457]. Plakortin (**243**) has moderate antiplasmodial activity, with an IC_{50} value of 0.87 μ*M* against the D10 strain and 0.39 μ*M* against the W2 strain, and the effects of stereochemistry on its antiplasmodial activity have been investigated [458]. Epiplakinidoic acid (**244**) from *P. halichondroides* had an IC_{50} value of 0.3 μ*M* against the W2 strain, but it also had significant cytotoxicity, and is thus not a viable lead compound [457].

9-Methoxystrobilurin G (**245**) is the most potent of a series of related compounds isolated from cultures of the basidiomycete *Favolaschia tonkinensis* [398, 399]. It had an IC_{50} value of 0.06 μ*M* against the K1 strain of *P. falciparum*, but only of 51 μ*M* against Vero cells, so it has good selectivity.

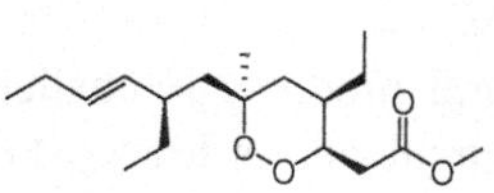

243 (plakortin)
D10 0.87 μ*M*
W2 0.41 μ*M*

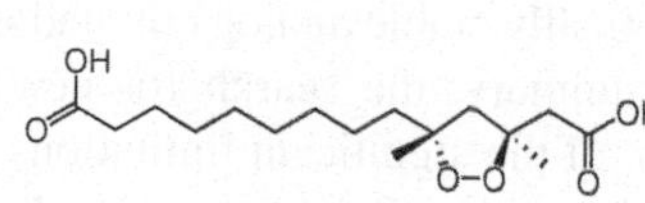

244 (epiplakinidioic acid)
W2 0.3 μ*M*
A2058 3 μ*M*
SI 10

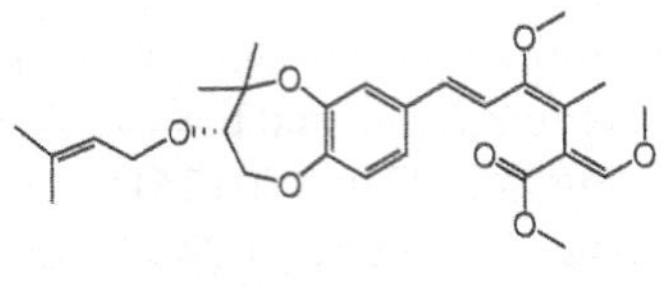

245 (9-methoxystrobilurin G)
K1 0.06 μ*M*
Vero 51 μ*M*

3.6 Conclusion

The compounds discussed in this contribution exemplify the successes and the challenges of the use of natural products to treat malaria. On the one hand, quinine and artemisinin and their analogs and derivatives, as well as the other clinically used compounds reviewed, demonstrate the enormous value of natural products in malaria treatment, and cumulatively have saved millions of lives. On the other hand, the necessarily strict limitations on the development of antimalarial drugs, and especially the need for low-cost therapeutic agents, means that newly discovered antiplasmodial compounds face a high hurdle for development. In spite of these limitations, the search for new natural products as antimalarial drugs or as lead compounds for the development of potent low-cost synthetic analogs continues to be productive.

Among the alkaloids discussed in this contribution the simple alkaloid cassiarin A (**52**) appears to have a favorable combination of synthetic accessibility, antiplasmodial activity, and low toxicity, so further SAR studies on it could prove fruitful. Dioncophylline C (**59**) and dioncopeltine A (**60**) both have in vivo activity against *P. berghei*, as does manzamine A (**103**). The febrifugine analogs **126–128** all have promising combinations of antiplasmodial activity with low mammalian toxicity, and are likely to reward further studies. Cipargamin (**106**), a synthetic alkaloid analog, is in clinical development, as is the alkaloid-like acridone (**55**) [263].

Terpenoids with development potential include ineupatorolide (**164**), with activity against *P. berghei* in mice, the quassinoids **179–181**, terpenoid-inspired synthetic isonitriles such as **159**, and fimbricalyx B (**169**). Other development candidates include the puberolic acid analogs **187** and **188**, cyclomarin A (**226**) and possibly metabolically stable analogs of cladosporin (**238**).

In summary, the search for new antimalarial natural products is challenging because of the significant limitations posed by the requirements for low-cost orally active drugs. In spite of these challenges, some new natural product-based drugs are in development, and the search for additional lead compounds from Nature's combinatorial library is very likely to yield new lead compounds or even drugs.

References

1. Arrow KJ, Panosian C, Gelband H (2004) A brief history of malaria. In: Arrow KJ, Panosian C, Gelband H (eds) Saving lives, buying time. Economics of malaria drugs in an age of resistance. National Academies Press, Washington, DC, p 125
2. Cunha CB, Cunha BA (2008) Brief history of the clinical diagnosis of malaria: from Hippocrates to Osler. J Vector Borne Dis 45:194
3. Jones WHS (1907) Malaria: a neglected factor in Greek history. With an introduction by Maj R Ross, FRS, CB and a concluding chapter by Ellett GG. Macmillan and Bowes, Cambridge, UK
4. Baron C, Hamlin C (2015) Malaria and the decline of ancient Greece: revisiting the Jones hypothesis in an era of interdisciplinarity. Minerva 53:327
5. Marciniak S, Prowse TL, Herring DA, Klunk J, Kuch M, Duggan AT, Bondioli L, Holmes EC, Poinar HN (2016) *Plasmodium falciparum* malaria in 1st–2nd century CE southern Italy. Curr Biol 26:R1220
6. Anonymous (2005) Diseases and causes of death among the Popes. Acta Theol Suppl 7:233
7. Bruce-Chwatt LJ (1977) John Macculloch, MD, FRS (1773–1835) (The precursor of the discipline of malariology). Med Hist 21:156
8. Gowland RL, Western AG (2012) Morbidity in the marshes: using spatial epidemiology to investigate skeletal evidence for malaria in Anglo-Saxon England (AD 410–1050). Am J Phys Anthropol 147:301
9. Kuhn KG, Campbell-Lendrum DH, Armstrong B Davies CR (2003) Malaria in Britain: past, present, and future. Proc Nat Acad Sci USA 100:9997
10. Mann CC (2011) 1493: Uncovering the New World Columbus created. Knopf, New York, p 560
11. Cox FEG (2010) History of the discovery of the malaria parasites and their vectors. Parasites Vectors 3:5
12. McCullough D (1977) The path between the seas: the creation of the Panama Canal, 1870–1914. Simon and Schuster, New York, p 698
13. Anonymous (2020) World malaria report 2020: 20 years of global progress and challenges. World Health Organization, Geneva, p 299
14. Caminade C, Kovats S, Rocklov J, Tompkins AM, Morse AP, Colón-González FJ, Stenlund H, Martens P Lloyd SJ (2014) Impact of climate change on global malaria distribution. Proc Natl Acad Sci USA 111:3286
15. Normile D (2021) It's official: China has eliminated malaria. Science. https://www.sciencemag.org/news/2021/06/it-s-official-china-has-eliminated-malaria
16. Constantinou K (1998) Anopheles (malaria) eradication in Cyprus. Parassitologia 40:131
17. Simac JN, Badar S, Farber JA, Brako MYO, Giudice-Jimenez RAL, Raspa SS, Achore M, MacKnight SD (2017) Malaria elimination in Sri Lanka. J Health Specialties 5:60
18. Anonymous (2021). WHO recommends groundbreaking malaria vaccine for children at risk https://www.who.int/news/item/06-10-2021-who-recommends-groundbreaking-malaria-vaccine-for-children-at-risk. Accessed 28 October 2021
19. Hsu E (2006) The history of Qing hao in the Chinese materia medica. Trans R Soc Trop Med Hyg 100:505
20. Tu Y (2016) Artemisinin—a gift from traditional Chinese medicine to the world (Nobel lecture). Angew Chem Int Ed 55:10210
21. Sutherland CJ, Tanomsing N, Nolder D, Oguike M, Jennison C, Pukrittayakamee S, Dolecek C, Hien TT, do Rosario VE, Arez AP, Pinto J, Michon P, Escalante AA, Nosten F, Burke M, Lee R, Blaze M, Otto TD, Barnwell JW, Pain A, Williams J, White NJ, Day NP, Snounou G, Lockhart PJ, Chiodini PL, Imwong M, Polley SD (2010) Two nonrecombining sympatric forms of the human malaria parasite *Plasmodium ovale* occur globally. J Infect Dis 201:1544
22. Cox FE (2010) History of the discovery of the malaria parasites and their vectors. Parasit Vectors 3:5

23. Adams JH, Mueller I (2017) The biology of *Plasmodium vivax*. Cold Spring Harb Perspect Med 7:a025585
24. Collins WE, Jeffery GM (2007) *Plasmodium malariae*: parasite and disease. Clin Microbiol Rev 20:579
25. Collins WE (2012) *Plasmodium knowlesi*: a malaria parasite of monkeys and humans. Annu Rev Entomol 57:107
26. Sutherland CJ (2016) Persistent parasitism: the adaptive biology of malariae and ovale malaria. Trends Parasitol 32:808
27. Cowman AF, Healer J, Marapana D, Marsh K (2016) Malaria: biology and disease. Cell 167:610
28. Srivastava A, Creek DJ, Evans KJ, De Souza D, Schofield L, Muller S, Barrett MP, McConville MJ, Waters AP (2015) Host reticulocytes provide metabolic reservoirs that can be exploited by malaria parasites. PLoS Pathog 11:e1004882
29. White NJ (2011) Determinants of relapse periodicity in *Plasmodium vivax* malaria. Malar J 10:297
30. Richter J, Franken G, Holtfreter MC, Walter S, Labisch A, Mehlhorn H (2016) Clinical implications of a gradual dormancy concept in malaria. Parasitol Res 115:2139
31. Trager W, Jensen JB (1976) Human malaria parasites in continuous culture. Science 193:673
32. Lee W-C, Russell B, Rénia L (2019) Sticking for a cause: the *falciparum* malaria parasites cytoadherence paradigm. Front Immunol 10:1444
33. Josling GA, Williamson KC, Llinás M (2018) Regulation of sexual commitment and gametocytogenesis in malaria parasites. Ann Rev Microbiol 72:501
34. Josling GA, Llinás M (2019) Commitment isn't for everyone. Trends Parasitol 35:381
35. Bancells C, Llorà-Batlle O, Poran A, Nötzel C, Rovira-Graells N, Elemento O, Kafsack BFC, Cortés A (2019) Revisiting the initial steps of sexual development in the malaria parasite *Plasmodium falciparum*. Nature Microbiol 4:144
36. Venugopal K, Hentzschel F, Valkiūnas G, Marti M (2020) *Plasmodium* asexual growth and sexual development in the haematopoietic niche of the host. Nature Rev Microbiol 18:177
37. De Niz M, Meibalan E, Mejia P, Ma S, Brancucci NMB, Agop-Nersesian C, Mandt R, Ngotho P, Hughes KR, Waters AP, Huttenhower C, Mitchell JR, Martinelli R, Frischknecht F, Seydel KB, Taylor T, Milner D, Heussler VT, Marti M (2018) *Plasmodium* gametocytes display homing and vascular transmigration in the host bone marrow. Science Adv 4:eaat3775
38. Nilsson SK, Childs LM, Buckee C, Marti M (2015) Targeting human transmission biology for malaria elimination. PLoS Pathog 11:e1004871
39. Laurens MB (2018) The promise of a malaria vaccine—are we closer? Ann Rev Microbiol 72:273
40. Bennink S, Kiesow MJ, Pradel G (2016) The development of malaria parasites in the mosquito midgut. Cell Microbiol 18:905
41. Billker O, Lindo V, Panico M, Etienne AE, Paxton T, Dell A, Rogers M, Sinden RE Morris HR (1998) Identification of xanthurenic acid as the putative inducer of malaria development in the mosquito. Nature 392:289
42. Sinden RE (1983) Sexual development of malarial parasites. In: Baker JR, Muller R (eds) Advances in parasitology, vol 22. Academic Press, London, New York, p 153
43. Sinden RE, Butcher GA, Billker O Fleck SL (1996) Regulation of infectivity of *Plasmodium* to the mosquito vector. In: Baker JR, Muller R, Rollinson D (eds) Advances in parasitology, vol 38. Academic Press, London, New York, p 53
44. Hayton K, Templeton TJ (2008) Osmiophilic bodies and the odd organelles of alveolates. Mol Microbiol 67:236
45. Paul REL, Brey PT, Robert V (2002) *Plasmodium* sex determination and transmission to mosquitoes. Trends Parasitol 18:32
46. Sinden RE, Carter R, Drakeley C, Leroy D (2012) The biology of sexual development of *Plasmodium*: the design and implementation of transmission-blocking strategies. Malaria J 11:70
47. Beier JC (1998) Malaria parasite development in mosquitoes. Ann Rev Entomol 43:519

48. Lang-Unnasch N, Murphy AD (1998) Metabolic changes of the malaria parasite during the transition from the human to the mosquito host. Ann Rev Microbiol 52:561
49. Ramasamy MS, Kulasekera R, Srikrishnaraj KA, Ramasamy R (1996) Different effects of modulation of mosquito (Diptera:Culicidae) trypsin activity on the infectivity of two human malaria (Hemosporidia:Plasmodidae) parasites. J Med Entomol 33:777
50. Meis JFGM, Pool G, Gemert GJ, Lensen AHW, Ponnudurai T, Meuwissen JHET (1989) *Plasmodium falciparum* ookinetes migrate intercellularly through *Anopheles stephensi* midgut epithelium. Parasitol Res 76:13
51. Greenwood D (1992) The quinine connection. J Antimicrob Chemother 30:417
52. Guerra F (1977) The introduction of *Cinchona* in the treatment of malaria. Part I. J Trop Med Hyg 80:112
53. Guerra F (1977) The introduction of *Cinchona* in the treatment of malaria. Pt. II. J Trop Med Hyg 80:135
54. Le Couteur P, Burreson J (2004) Napoleon's buttons: 17 molecules that changed history. Tarcher/Penguin, New York, p 375
55. Gannon M (2015) What really killed notorious English leader Oliver Cromwell? https://www.livescience.com/52569-oliver-cromwell-death-typhoid-fever.html. Accessed 14 Oct 2019
56. Kaufman TS, Ruveda EA (2005) The quest for quinine: those who won the battles and those who won the war. Angew Chem Int Ed Engl 44:854
57. Cova TFGG, Pais AACC, Seixas de Melo JS (2017) Reconstructing the historical synthesis of mauveine from Perkin and Caro: procedure and details. Sci Rep 7:6806
58. Bozic B, Uzelac TV, Kezic A Bajcetic M (2018) The role of quinidine in the pharmacological therapy of ventricular arrhythmias 'quinidine'. Mini Rev Med Chem 18:468
59. Yang F, Hanon S, Lam P, Schweitzer P (2009) Quinidine revisited. Am J Med 122:317
60. Leete E (1960) Biosynthesis of quinine and related alkaloids. Acc Chem Res 2:59
61. Battersby AR, Hall ES (1970) Biosynthesis of quinine from loganin. J Chem Soc D:194
62. Battersby AR, Parry RJ (1971) Biosynthesis of the *Cinchona* alkaloids: late stages of the pathway. J Chem Soc D:31
63. Battersby AR Parry RJ (1971) Biosynthesis of the *Cinchona* alkaloids: middle stages of the pathway. J Chem Soc D:30
64. O'Connor SE, Maresh JJ (2006) Chemistry and biology of monoterpene indole alkaloid biosynthesis. Nat Prod Rep 23:532
65. Stevens LH, Giroud C, Pennings EJM, Verpoorte R (1993) Purification and chacterization of strictosidine synthase from a suspension culture of *Cinchona robusta*. Phytochemistry 33:99
66. Trenti F, Yamamoto K, Hong B, Paetz C, Nakamura Y, O'Connor SE (2021) Early and late steps of quinine biosynthesis. Org Lett 23:1793
67. de Villiers KA, Gildenhuys J, le Roex T (2012) Iron(III) protoporphyrin IX complexes of the antimalarial *Cinchona* alkaloids quinine and quinidine. ACS Chem Biol 7:666
68. Alumasa JN, Gorka AP, Casabianca LB, Comstock E, de Dios AC, Roepe PD (2011) The hydroxyl functionality and a rigid proximal N are required for forming a novel non-covalent quinine-heme complex. J Inorg Biochem 105:467
69. Olafson KN, Nguyen TQ, Rimer JD, Vekilov PG (2017) Antimalarials inhibit hematin crystallization by unique drug-surface site interactions. Proc Natl Acad Sci USA 114:7531
70. White NJ (1998) Drug resistance in malaria. Brit Med Bull 54:703
71. Achan J, Talisuna AO, Erhart A, Yeka A, Tibenderana JK, Baliraine FN, Rosenthal PJ, D'Alessandro U (2011) Quinine, an old anti-malarial drug in a modern world: role in the treatment of malaria. Malaria J 10:144
72. Le Bras J, Durand R (2003) The mechanisms of resistance to antimalarial drugs in *Plasmodium falciparum*. Fund Clin Pharmacol 17:147
73. Conrad MD, Rosenthal PJ (2019) Antimalarial drug resistance in Africa: the calm before the storm? Lancet Infect Dis 19:e338
74. Phyo AP, Win KK, Thu AM, Swe LL, Htike H, Beau C, Sriprawat K, Winterberg M, Proux S, Imwong M, Ashley EA, Nosten F (2018) Poor response to artesunate treatment in two patients with severe malaria on the Thai-Myanmar border. Malar J 17:30

75. Greenwood D (1995) Historical perspective. Conflicts of interest: the genesis of synthetic antimalarial agents in peace and war. J Antimicrob Chemother 36:857
76. Foley M, Tilley L (1998) Quinoline antimalarials: mechanisms of action and resistance and prospects for new agents. Pharmacol Ther 79:55
77. Sullivan DJ Jr (2017) Quinolines block every step of malaria heme crystal growth. Proc Natl Acad Sci USA 114:7483
78. Muehlens P (1926) Die Behandlung der natuerlichen menschlichen Malaria-Infektionen mit Plasmochin. Arch Schiffs-u Tropenhyg 30:25
79. Nanayakkara NPN, Tekwani BL, Herath HM, Sahu R, Gettayacamin M, Tungtaeng A, van Gessel Y, Baresel P, Wickham KS, Bartlett MS, Fronczek FR, Melendez V, Ohrt C, Reichard GA, McChesney JD, Rochford R, Walker LA (2014) Scalable preparation and differential pharmacologic and toxicologic profiles of primaquine enantiomers. Antimicrob Agents Chemother 58:4737
80. Vale N, Moreira R Gomes P (2009) Primaquine revisited six decades after its discovery. Eur J Med Chem 44:937
81. Coatney CR (1961) Pitfalls in a discovery: the chronicle of chloroquine. Am J Trop Med Hyg 1963:121
82. Anonymous (2019) Mepacrine. https://en.wikipedia.org/wiki/Mepacrine. Accessed 18 Oct 2019
83. Loeb RF, Clarke WM, Coateney GR, Coggeshall LT, Dieuaide FR, Dochez AR, Hakansson EG, Marshall EK, Marvel SC, McCoy OR, Sapero JJ, Serbell WH, Shannon JA, Carden GA (1946) Activity of a new antimalarial agent, chloroquine (SN 7618). J Am Med Assoc 130:1069
84. Slater AFG (1993) Chloroquine: mechanism of drug action and resistance in *Plasmodium falciparum*. Pharmacol Ther 57:203
85. Paynre D (1988) Did medicated salt hasten the spread of chloroquine resistance in *Plasmodium falciparum*? Parasitol Today 4:112
86. Kitagawa T, Matsumoto A, Terashima I, Uesono Y (2021) Antimalarial quinacrine and chloroquine lose their activity by decreasing cationic amphiphilic structure with a slight decrease in pH. J Med Chem 64:3885
87. Croft AM (2007) A lesson learnt: the rise and fall of Lariam and Halfan. J R Soc Med 100:170
88. Tan KR, Hwang J (2018) Tafenoquine receives regulatory approval in USA for prophylaxis of malaria and radical cure of *Plasmodium vivax*. J Trav Med 25:1
89. Wiesner J, Ortmann R, Jomaa H Schlitzer M (2003) New antimalarial drugs. Angew Chem Int Ed 42:5274
90. Talapko J, Skrlec I, Alebic T, Jukic M, Vcev A (2019) Malaria: the past and the present. Microorganisms 7:179
91. Pybus BS (2016) Synthetic vs. natural bioactive compounds against tropical disease. In: Vale N (ed) Biomedical chemistry current trends and developments. De Gruyter, Berlin, p 275
92. Guiguemde WA, Shelat AA, Bouck D, Duffy S, Crowther GJ, Davis PH, Smithson DC, Connelly M, Clark J, Zhu F, Jimenez-Diaz MB, Martinez MS, Wilson EB, Tripathi AK, Gut J, Sharlow ER, Bathurst I, El Mazouni F, Fowble JW, Forquer I, McGinley PL, Castro S, Angulo-Barturen I, Ferrer S, Rosenthal PJ, DeRisi JL, Sullivan DJ Jr, Lazo JS, Roos DS, Riscoe MK, Phillips MA, Rathod PK, Voorhis WCV, Avery VM, Guy RK (2010) Chemical genetics of *Plasmodium falciparum*. Nature 465:311
93. Gamo F-J, Sanz LM, Vidal J, de Cozar C, Alvarez E, Lavandera J-L, Vanderwall DE, Green DVS, Kumar V, Hasan S, Brown JR, Peishoff CE, Cardon LR, Garcia-Bustos JF (2010) Thousands of chemical starting points for antimalarial lead identification. Nature 465:305
94. Narula AK, Azad CS, Nainwal LM (2019) New dimensions in the field of antimalarial research against malaria resurgence. Eur J Med Chem 181:111353
95. Brown GD (2010) The biosynthesis of artemisinin (Qinghaosu) and the phytochemistry of *Artemisia annua* L. (Qinghao). Molecules 15:7603
96. Hsu E (2006) Reflections on the 'discovery' of the antimalarial qinghao. Br J Clin Pharmacol 61:666

97. Tu Y (2011) The discovery of artemisinin (qinghaosu) and gifts from Chinese medicine. Nat Med 17:1217
98. Yearsley C (2016) Artemisinin: a Nobel-winning antimalarial from traditional Chinese medicine. HerbalGram 110:50
99. Kong LY, Tan RX (2015) Artemisinin, a miracle of traditional Chinese medicine. Nat Prod Rep 32:1617
100. Efferth T (2007) Willmar Schwabe Award 2006: Antiplasmodial and antitumor activity of artemisinin: from bench to bedside. Planta Med 73:299
101. Meshnick SR, Taylor TE Kamchonwongpaisan S (1996) Artemisinin and the antimalarial endoperoxides: from herbal remedy to targeted chemotherapy. Microbiol Rev 60:301
102. Liu C-X (2017) Discovery and development of artemisinin and related compounds. Chin Herbal Med 9:101
103. Anonymous (1977) A new type of sesquiterpene lactone—Qing Hau Sau. Kexue Tongbao (Chinese) 22:142
104. Klayman DL, Lin AJ, Acton N, Scovill JP, Hoch JM, Milhous WK, Theoharides AD, Dobek AS (1984) Isolation of artemisinin (Qinghaosu) from *Artemisia annua* growing in the United States. J Nat Prod 47:715
105. Klayman DL (1985) *Qinghaosu* (artemisinin): an antimalarial drug from China. Science 228:1049
106. Bhakuni RS, Jain DC, Sharma RP (2002) Phytochemistry of *Artemisia annua* and the development of artemisinin-derived antimalarial agents. In: Wright CW (ed) *Artemisia*. Taylor and Francis, Oxford, UK, p 211
107. Bhakuni RS, Jain DC, Sharma RP Kumar S (2001) Secondary metabolites of *Artemisia annua* and their biological activity. Curr Sci 80:35
108. Sangwan NS, Kumar R, Srivastava S, Kumar A, Gupta A, Sangwan RS (2010) Recent developments on secondary metabolite biosynthesis in *Artemisia annua* L. J Plant Biol 37:1
109. Ikram N, Simonsen HT (2017) A review of biotechnological artemisinin production in plants. Front Plant Sci 8:1966
110. Bouwmeester HJ, Wallaart TEE, Janssen MHA, Loo BV, Jansen BJM, Posthumus MA, Schmidt CO, Kraker JD, Kong WA, Franssen MCR (1999) Amorpha-4,11-diene synthase catalyses the first probable step in artemisinin synthesis. Phytochemistry 52:843
111. Akhila A, Rabni K, Thakur RS (1990) Biosynthesis of artemisinic acid in *Artemisia annua*. Phytochemistry 29:2129
112. Akhila A, Thakur RS, Popli SP (1987) Biosyntheis of artemisinin in *Artemisia annua*. Phytochemistry 26:1927
113. Schramek N, Wang H, Romisch-Margl W, Keil B, Radykewicz T, Winzenhorlein B, Beerhues L, Bacher A, Rohdich F, Gershenzon J, Liu B, Eisenreich W (2010) Artemisinin biosynthesis in growing plants of *Artemisia annua*. A $^{13}CO_2$ study. Phytochemistry 71:179
114. Srivastava N, Akhila A (2011) Biosynthesis of artemisinin—revisited. J Plant Interact 6:265
115. Czechowski T, Larson TR, Catania TM, Harvey D, Brown GD Graham IA (2016) *Artemisia annua* mutant impaired in artemisinin synthesis demonstrates importance of nonenzymatic conversion in terpenoid metabolism. Proc Natl Acad Sci USA 113:15150
116. Peplow M (2018) Looking for cheaper routes to malaria medicines. Chem Eng News 96:29
117. Cook S (2014) Artemisinin: a case study in the evolution of synthetic strategy. Synlett 25:751
118. Corsello MA, Garg NK (2015) Synthetic chemistry fuels interdisciplinary approaches to the production of artemisinin. Nat Prod Rep 32:359
119. Schmid G, Hofheinz W (1983) Total synthesis of qinghaosu. J Am Chem Soc 105:624
120. Zhou W-S (1986) Total synthesis of arteannuin (Quinghaosu) and related compounds. Pure Appl Chem 58:817
121. Xu X-X, Shu J, Huang D-Z, Zhou WS (1986) Total synthesis of artenannuin and deoxyarteannuin. Tetrahedron 42:819
122. Zhou W-S, Xu X-X (1994) Total synthesis of the antimalarial sesquiterpene peroxide qinghaosu and yingzhaosu A. Acc Chem Res 27:211

123. Ravindranathan T, Kumar MA, Menon RB, Hiremath SV (1990) Stereoselective synhesis of artemisinin. Tetrahedron Lett 31:755
124. Bhonsle JB, Pandey B, Deshpande VH, Ravindranathan T (1994) New synthetic strategies towards (+)-artemisinin. Tetrahedron Lett 35:5489
125. Liu H-J, Yeh WL, Chew SY (1993) A total synthesis of the antimalarial natural product (+)-qinghaosu. Tetrahedron Lett 34:4435
126. Avery MA, Chong WKM, Jennings-White C (1992) Stereoselective total synthesis of (+)-artemisinin, the antimalarial constituent of *Artemisia annua* L. J Am Chem Soc 114:974
127. Yadav JS, Babu RS, Sabitha G (2003) Stereoselective total synthesis of (+)-artemisinin. Tetrahedron Lett 44:387
128. Yadav JS, Thirupathaiah B, Srihari P (2010) A concise stereoselective total synthesis of (+)-artemisinin. Tetrahedron 66:2005
129. Zhu C, Cook S (2012) A concise synthesis of (+)-artemisinin. J Am Chem Soc 134:1377
130. Krieger J, Smeilus T, Kaiser M, Seo EJ, Efferth T Giannis A (2018) Total synthesis and biological investigation of (–)-artemisinin: the antimalarial activity of artemisinin is not stereospecific. Angew Chem Int Ed Engl 57:8293
131. Li ZL, Gu HM, Warhurst DC, Peters W (1983) Effects of qinghaosu and related compounds on incorporation of [G-^{3}H] hypoxanthine by *Plasmodium falciparum* in vitro. Trans Roy Soc Trop Med Hyg 77:522
132. Kumari A, Karnatak M, Singh D, Shankar R, Jat JL, Sharma S, Yadav D, Shrivastava R, Verma VP (2019) Current scenario of artemisinin and its analogues for antimalarial activity. Eur J Med Chem 163:804
133. Xu L (2017) Artesunate. In: Li G, Li Y, Li Z, Zeng M, Arnold M, Arnold K (eds) Artemisinin-based and other antimalarials: detailed account of studies by Chinese scientists who discovered and developed them. Academic Press, London, p 20
134. Magri NF, Kingston DGI, Jitrangsri C, Piccariello T (1986) Modified taxols. 3. Preparation and acylation of baccatin III. J Org Chem 51:3239
135. Presser A, Feichtinger A, Buzzi S (2017) A simplified and scalable synthesis of artesunate. Monatsh Chem 148:63
136. Gilmore K, Kopetzki D, Lee JW, Horvath Z, McQuade DT, Seidel-Morgenstern A, Seeberger PH (2014) Continuous synthesis of artemisinin-derived medicines. Chem Commun 50:12652
137. Luo X-D, Shen C-C (1987) The chemistry, pharmacology, and clinical applications of Qinghaosu (artemisinin) and its derivatives. Med Res Rev 7:29
138. Jeyadevan JP, Bray PG, Chadwick J, Mercer AE, Byrne A, Ward SA, Park BK, Dominic PW, Cosstick R, Davies JD, Higson AP, Irving E, Posner GH, O'Neill PM (2004) Antimalarial and antitumor evaluation of novel C-10 non-acetal dimers of 10β-(2-hydroxyethyl)deoxoartemisinin. J Med Chem 47:1290
139. Posner GH, Paik I-H, Chang W, Borstnik K, Sinishtaj S, Rosenthal AS, Shapiro TA (2007) Malaria-infected mice are cured by a single dose of novel artemisinin derivatives. J Med Chem 50:2516
140. Moon DK, Tripathi A, Siegler MA, Parkin S, Posner GH (2011) A single, low, oral dose of a 5-carbon-linked trioxane dimer orthoester plus mefloquine cures malaria-infected mice. Bioorg Med Chem Lett 21:2773
141. Jacobine AM, Mazzone JR, Slack RD, Tripathi AK, Sullivan DJ, Posner GH (2012) Malaria-infected mice live until at least day 30 after a new artemisinin-derived thioacetal thiocarbonate combined with mefloquine are administered together in a single, low, oral dose. J Med Chem 55:7892
142. Ferreira LT, Borba JVB, Moreira-Filho JT, Rimoldi A, Andrade CH, Costa FTM (2021) QSAR-based virtual screening of natural products database for identification of potent antimalarial hits. Biomolecules 11:459
143. Vennerstrom JL, Arbe-Barnes S, Brun R, Charman SA, Chiu FCK, Chollet J, Dong Y, Dorn A, Hunziker D, Matile H, McIntosh K, Padmanilayam M, Tomas JS, Scheurer C, Scorneaux B, Tang Y, Urwyler H, Wittlin S, Charman WN (2004) Identification of an antimalarial synthetic trioxolane drug development candidate. Nature 430:900

144. Rathi A (2012) Ranbaxy launches new anti-malarial Synriam. Chemistry World, RSC, London, May 2
145. Anonymous (2014) Ranbaxy receives approval for malaria drug Synriam from 7 African countries. Business Standard, New Delhi, December 17
146. Charman SA, Arbe-Barnes S, Bathurst IC, Brun R, Campbell M, Charman WN, Chiu FC, Chollet J, Craft JC, Creek DJ, Dong Y, Matile H, Maurer M, Morizzi J, Nguyen T, Papastogiannidis P, Scheurer C, Shackleford DM, Sriraghavan K, Stingelin L, Tang Y, Urwyler H, Wang X, White KL, Wittlin S, Zhou L, Vennerstrom JL (2011) Synthetic ozonide drug candidate OZ439 offers new hope for a single-dose cure of uncomplicated malaria. Proc Natl Acad Sci USA 108:4400
147. Phyo AP, Jittamala P, Nosten FH, Pukrittayakamee S, Imwong M, White NJ, Duparc S, Macintyre F, Baker M, Möhrle JJ (2016) Antimalarial activity of artefenomel (OZ439), a novel synthetic antimalarial endoperoxide, in patients with *Plasmodium falciparum* and *Plasmodium vivax* malaria: an open-label phase 2 trial. The Lancet Infect Dis 16:61
148. Anonymous (2020) Artefenomel. https://www.mmv.org/related-story-type/artefenomel. Accessed 20 Nov 2020
149. Salim M, Khan J, Ramirez G, Murshed M, Clulow AJ, Hawley A, Ramachandruni H, Beilles S, Boyd BJ (2019) Impact of ferroquine on the solubilization of artefenomel (OZ439) during in vitro lipolysis in milk and implications for oral combination therapy for malaria. Mol Pharmaceutics 16:1658
150. Cragg GM, Schepartz SA, Suffness M, Grever MR (1993) The taxol supply crisis. New NCI policies for handling the large scale production of novel natural product anticancer and anti-HIV agents. J Nat Prod 56:1657
151. Farhi M, Kozin M, Duchin S, Vainstein A (2013) Metabolic engineering of plants for artemisinin synthesis. Biotechnol Genet Eng Rev 29:135
152. Kopetzki D, Levesque F, Seeberger PH (2013) A continuous-flow process for the synthesis of artemisinin. Chem Eur J 19:5450
153. Levesque F, Seeberger PH (2012) Continuous-flow synthesis of the anti-malaria drug artemisinin. Angew Chem Int Ed Engl 51:1706
154. Holton RA, Biediger RJ, Boatman PD (1995) Semisynthesis of Taxol® and Taxotere. In: Suffness M (ed) Taxol: science and applications. CRC Press Inc., Boca Raton, FL, p 97
155. Wink M, Alfermann AW, Franke R, Wetterauer B, Distl M, Windhovel J, Krohn O, Fuss E, Garden H, Mohagheghzadeh A, Wildi E Ripplinger P (2005) Sustainable bioproduction of phytochemicals by plant in vitro cultures: anticancer agents. Plant Genet Resour 3:90
156. Woerdenbag HJ, Lüers JFJ, van Uden W, Pras N, Malingré TM Alfermann AW (1993) Production of the new antimalarial drug artemisinin in shoot cultures of *Artemisia annua*. Plant Cell Tissue Organ Cult 32:247
157. Baldi A, Dixit VK (2008) Yield enhancement strategies for artemisinin production by suspension cultures of *Artemisia annua*. Bioresource Technol 99:4609
158. Anonymous (2017) Phyton Biotech wins grant to develop new approaches to manufacturing artemisinin. https://phytonbiotech.com/phyton-biotech-wins-grant-develop-new-approaches-manufacturing-artemisinin-2/. Accessed 22 June 2021
159. Liu C-Z, Guo C, Wang Y-C, Ouyang F (2002) Effect of light irradiation on hairy root growth and artemisinin biosynthesis of *Artemisia annua* L. Proc Biochem 38:581
160. Patra N, Srivastava AK (2014) Mass scale artemisinin production in a stirred tank bioreactor using hairy roots of *Artemisia annua*. Int J Biosci Biochem Bioinform 4:467
161. Patra N, Srivastava A (2016) Artemisinin production by plant hairy root cultures in gas- and liquid-phase bioreactors. Plant Cell Rep 35:143
162. Patra N, Srivastava AK (2018) Mass production of artemisinin using hairy root cultivation of *Artemisia annua* in bioreactor. In: Pavlov A, Bley T (eds) Bioprocessing of plants in vitro systems. Springer Int Publ AG, Basel, Switzerland, p 343
163. Martin VJJ, Pitera DJ, Withers ST, Newman JD, Keasling JD (2003) Engineering a mevalonate pathway in *Escherichia coli* for production of terpenoids. Nature Biotechnol 21:796

164. Westfall PJ, Pitera DJ, Lenihan JR, Eng D, Woolard FX, Regentin R, Horning T, Tsuruta H, Melis DJ, Owens A, Fickes S, Diola D, Benjamin KR, Keasling JD, Leavell MD, McPhee DJ, Renninger NS, Newman JD, Paddon CJ (2012) Production of amorphadiene in yeast, and its conversion to dihydroartemisinic acid, precursor to the antimalarial agent artemisinin. Proc Natl Acad Sci USA 109:E111
165. Paddon CJ, Westfall PJ, Pitera DJ, Benjamin K, Fisher K, McPhee D, Leavell MD, Tai A, Main A, Eng D, Polichuk DR, Teoh KH, Reed DW, Treynor T, Lenihan J, Fleck M, Bajad S, Dang G, Dengrove D, Diola D, Dorin G, Ellens KW, Fickes S, Galazzo J, Gaucher SP, Geistlinger T, Henry R, Hepp M, Horning T, Iqbal T, Jiang H, Kizer L, Lieu B, Melis D, Moss N, Regentin R, Secrest S, Tsuruta H, Vazquez R, Westblade LF, Xu L, Yu M, Zhang Y, Zhao L, Lievense J, Covello PS, Keasling JD, Reiling KK, Renninger NS Newman JD (2013) High-level semi-synthetic production of the potent antimalarial artemisinin. Nature 496:528
166. Paddon CJ, Keasling JD (2014) Semi-synthetic artemisinin: a model for the use of synthetic biology in pharmaceutical development. Nat Rev Microbiol 12:355
167. Ro D-K, Paradise EM, Ouellet M, Fisher KJ, Newman KL, Ndungu JM, Ho KA, Eachus RA, Ham TS, Kirby J, Chang MCY, Withers ST, Shiba Y, Sarpong R, Keasling JD (2006) Production of the antimalarial drug precursor artemisinic acid in engineered yeast. Nature 440:940
168. Turconi J, Griolet F, Guevel R, Oddon G, Villa R, Geatti A, Hvala M, Rossen K, Göller R, Burgard A (2014) Semisynthetic artemisinin, the chemical path to industrial production. Org Proc Res Dev 18:417
169. Singh D, McPhee D, Paddon CJ, Cherry J, Maurya G, Mahale G, Patel Y, Kumar N, Singh S, Sharma B, Kushwaha L, Singh S, Kumar A (2017) Amalgamation of synthetic biology and chemistry for high-throughput nonconventional synthesis of the antimalarial drug artemisinin. Org Proc Res Dev 21:551
170. Chen H-J, Han W-B, Hao H-D, Wu Y (2013) A facile and scalable synthesis of qinghaosu (artemisinin). Tetrahedron 69:1112
171. Haynes RK, Vonwiller SC (1997) From Qinghao, marvelous herb of antiquity, to the antimalarial trioxane qinghaosu—and some remarkable new chemistry. Acc Chem Res 30:73
172. Tang X, Demiray M, Wirth T Allemann RK (2018) Concise synthesis of artemisinin from a farnesyl diphosphate analogue. Bioorg Med Chem 26:1314
173. Parayil A (2019) Manus Bio awarded next stage of funding to continue development of a novel production method for artemisinin, May 9, 2019. Global Newswire, Cambridge, MA
174. Covello PS (2008) Making artemisinin. Phytochemistry 69:2881
175. Kung SH, Lund S, Murarka A, McPhee D, Paddon CJ (2018) Approaches and recent developments for the commercial production of semi-synthetic artemisinin. Front Plant Sci 9:87
176. Brossi A, Venugopalan B, Dominguez Gerpe L, Yeh HJC, Flippen-Anderson JL, Buchs P, Luo XD, Milhous W, Peters W (1988) Arteether, a new antimalarial drug: synthesis and antimalarial properties. J Med Chem 31:645
177. Krungkrai SR, Yuthavong Y (1987) The antimalarial action on *Plasmodium falciparum* of qinghaosu and artesunate in combination with agents which modulate oxidant stress. Trans R Soc Trop Med Hyg 81:710
178. Klonis N, Creek DJ, Tilley L (2013) Iron and heme metabolism in *Plasmodium falciparum* and the mechanism of action of artemisinins. Curr Opin Microbiol 16:722
179. Klonis N, Crespo-Ortiz MP, Bottova I, Abu-Bakar N, Kenny S, Rosenthal PJ, Tilley L (2011) Artemisinin activity against *Plasmodium falciparum* requires hemoglobin uptake and digestion. Proc Natl Acad Sci USA 108:11405
180. Meshnick SR (2001) Artemisinin and its derivatives. In: Rosenthal PJ (ed) Antimalarial chemotherapy. Humana Press, Totowa, NJ, p 191
181. O'Neill PM, Posner GH (2004) A medicinal chemistry perspective on artemisinin and related endoperoxides. J Med Chem 47:2945
182. Haynes RK, Cheu K-W, N'Da D, Coghi P, Monti D (2013) Considerations on the mechanism of action of artemisinin antimalarials: Part 1—the 'carbon radical' and 'heme' hypotheses. Infect Disord Drug Targets 13:2117

183. Meunier B, Robert A (2010) Heme as trigger and target for trioxane-containing antimalarial drugs. Acc Chem Res 43:1444
184. Berman PA, Adams PA (1997) Artemisinin enhances heme-catalysed oxidation of lipid membranes. Free Radical Biol Med 22:1283
185. Wu Y (2002) How might qinghaosu (artemisinin) and related compounds kill the intraerythrocytic malaria parasite? A chemist's view. Acc Chem Res 35:255
186. Haynes RK, Ho WY, Chan HW, Fugmann B, Stetter J, Croft SL, Vivas L, Peters W, Robinson BL (2004) Highly antimalaria-active artemisinin derivatives: biological activity does not correlate with chemical reactivity. Angew Chem Int Ed Engl 43:1381
187. Haynes RK, Chan WC, Lung CM, Uhlemann AC, Eckstein U, Taramelli D, Parapini S, Monti D, Krishna S (2007) The Fe^{2+}-mediated decomposition, PfATP6 binding and antimalarial activities of artemisone and other artemisinins: the unlikelihood of C-centered radicals as bioactive intermediates. ChemMedChem 2:1480
188. Pal C, Sarkar S, Mazumder S, Adhikari S, Bandyopadhyay U (2013) Synthesis and biological evaluation of primaquine–chloroquine twin drug: a novel heme-interacting molecule prevents free heme and hydroxyl radical-mediated protein degradation. MedChemComm 4:731
189. Eckstein-Ludwig U, Webb RJ, van Goethem IDA, East JM, Lee AG, Kimura M, O'Neill PM, Bray PG, Ward SA, Krishna S (2003) Artemisinins target the SERCA of *Plasmodium falciparum*. Nature 424:957
190. del Pilar CM, Avery TD, Hanssen E, Fox E, Robinson TV, Valente P, Taylor DK, Tilley L (2008) Artemisinin and a series of novel endoperoxide antimalarials exert early effects on digestive vacuole morphology. Antimicrob Agents Chemother 52:98
191. Garah FB, Stigliani JL, Cosledan F, Meunier B, Robert A (2009) Docking studies of structurally diverse antimalarial drugs targeting PfATP6: no correlation between in silico binding affinity and in vitro antimalarial activity. ChemMedChem 4:1469
192. Asawamahasakda W, Ittarat I, Pu YM, Ziffer H, Meshnick SR (1994) Reaction of antimalarial endoperoxides with specific parasite proteins. Antimicrob Agents Chemother 38:1854
193. Wang J, Zhang C-J, Chia WN, Loh CC, Li Z, Lee YM, He Y, Yuan LX, Lim TK, Liu M, Liew CX, Lee YQ, Zhang J, Lu N, Lim CT, Hua ZC, Liu B, Shen HM, Tan KS, Lin Q (2015) Haem-activated promiscuous targeting of artemisinin in *Plasmodium falciparum*. Nat Commun 6:10111
194. Ismail HM, Barton V, Phanchana M, Charoensutthivarakul S, Wong MH, Hemingway J, Biagini GA, O'Neill PM, Ward SA (2016) Artemisinin activity-based probes identify multiple molecular targets within the asexual stage of the malaria parasites *Plasmodium falciparum* 3D7. Proc Natl Acad Sci USA 113:2080
195. Ismail HM, Barton VE, Panchana M, Charoensutthivarakul S, Biagini GA, Ward SA, O'Neill PM (2016) A click chemistry-based proteomic approach reveals that 1,2,4-trioxolane and artemisinin antimalarials share a common protein alkylation profile. Angew Chem Int Ed Engl 55:6401
196. Haynes RK, Cheu KW, Chan HW, Wong HN, Li KY, Tang MM, Chen MJ, Guo ZF, Guo ZH, Sinniah K, Witte AB, Coghi P, Monti D (2012) Interactions between artemisinins and other antimalarial drugs in relation to the cofactor model—a unifying proposal for drug action. ChemMedChem 7:2204
197. Efferth T (2017) Cancer combination therapies with artemisinin-type drugs. Biochem Pharmacol 139:56
198. Efferth T (2017) From ancient herb to modern drug: *Artemisia annua* and artemisinin for cancer therapy. Semin Cancer Biol 46:65
199. Taleghani A, Emami SA, Tayarani-Najaran Z (2020) *Artemisia*: a promising plant for the treatment of cancer. Bioorg Med Chem 28:115180
200. Mancuso RI, Foglio MA, Olalla Saad ST (2021) Artemisinin-type drugs for the treatment of hematological malignancies. Cancer Chemother Pharmacol 87:1
201. Efferth T (2018) Beyond malaria: the inhibition of viruses by artemisinin-type compounds. Biotechnol Adv 36:1730

202. Wu G, Cheng B, Qian H, Ma S, Chen Q (2019) Identification of HSP90 as a direct target of artemisinin for its anti-inflammatory activity via quantitative chemical proteomics. Org Biomol Chem 17:6854
203. Peplow M (2020) Artemisinin raises hopes and fears amid COVID-19. Chem Eng News 98, May 27
204. Pepe DA, Toumpa D, André-Barrès C, Menendez C, Mouray E, Baltas M, Grellier P, Papaioannou D, Athanassopoulos CM (2020) Synthesis of novel G factor or chloroquine-artemisinin hybrids and conjugates with potent antiplasmodial activity. ACS Med Chem Lett 11:921
205. Dondorp AM, Nosten F, Yi P, Das D, Phyo AP, Tarning J, Lwin KM, Ariey F, Hanpithakpong W, Lee SJ, Ringwald P, Silamut K, Imwong M, Chotivanich K, Lim P, Herdman T, An SS, Yeung S, Singhasivanon P, Day NP, Lindegardh N, Socheat D, White NJ (2009) Artemisinin resistance in *Plasmodium falciparum* malaria. New Engl J Med 361:455
206. Dondorp AM, Fairhurst RM, Slutsker L, MacArthur JR, Breman JG, Guerin PJ, Wellems TE, Ringwald P, Newman RD, Plowe CV (2011) The threat of artemisinin-resistant malaria. N Engl J Med 365:1073
207. Fairhurst RM, Dondorp AM (2016) Artemisinin-resistant *Plasmodium falciparum* malaria. Microbiol. Spectrum 4:36
208. Birnbaum J, Scharf S, Schmidt S, Jonscher E, Hoeijmakers WAM, Flemming S, Toenhake CG, Schmitt M, Sabitzki R, Bergmann B, Fröhlke U, Mesén-Ramírez P, Soares AB, Herrmann H, Bártfai R, Spielmann T (2020) A Kelch13-defined endocytosis pathway mediates artemisinin resistance in malaria parasites. Science 367:51
209. Suresh N, Haldar K (2018) Mechanisms of artemisinin resistance in *Plasmodium falciparum* malaria. Curr Opin Pharmacol 42:46
210. Rosenthal MR, Ng CL (2020) *Plasmodium falciparum* artemisinin resistance: the effect of heme, protein damage, and parasite cell stress response. ACS Infect Dis 6:1599
211. Gaillard T, Madamet M, Pradines B (2015) Tetracyclines in malaria. Malar J 14:445
212. Roberts MC (1996) Tetracycline resistance determinants: mechanisms of action, regulation of expression, genetic mobility, and distribution. FEMS Microbiol Rev 19:1
213. Dahl EL, Shock JL, Shenai BR, Gut J, DeRisi JL, Rosenthal PJ (2006) Tetracyclines specifically target the apicoplast of the malaria parasite *Plasmodium falciparum*. Antimicrob Agents Chemother 50:3124
214. Tan KR, Magill AJ, Parise ME, Arguin PM (2011) Doxycycline for malaria chemoprophylaxis and treatment: report from the CDC expert meeting on malaria chemoprophylaxis. Am J Trop Med Hyg 84:517
215. Spizek J, Rezanka T (2004) Lincomycin, clindamycin and their applications. Appl Microbiol Biotechnol 64:455
216. Fichera ME, Roos DS (1997) A plastid organelle as a drug target in apicomplexan parasites. Nature 390:407
217. Lell B, Kremsner PG (2002) Clindamycin as an antimalarial drug: review of clinical trials. Antimicrob Agents Chemother 46:2315
218. Saito M, Gilder ME, McGready R, Nosten F (2018) Antimalarial drugs for treating and preventing malaria in pregnant and lactating women. Expert Opin Drug Saf 17:1129
219. Bakheit AH, Al-Hadiya BM, Abd-Elgalil AA (2014) Azithromycin. Profiles of drug substances, excipients, and related methodology. Elsevier, Amsterdam, p 1
220. Gaillard T, Dormoi J, Madamet M, Pradines B (2016) Macrolides and associated antibiotics based on similar mechanism of action like lincosamides in malaria. Malar J 15:85
221. van Eijk AM, Terlouw DJ (2011) Azithromycin for treating uncomplicated malaria. Cochrane Database Syst Rev 2011:CD006688
222. Burns AL, Sleebs BE, Siddiqui G, De Paoli AE, Anderson D, Liffner B, Harvey R, Beeson JG, Creek DJ, Goodman CD, McFadden GI, Wilson DW (2020) Retargeting azithromycin analogues to have dual-modality antimalarial activity. BMC Biol 18:133
223. Kuroda Y, Okuhara M, Goto T, Okamoto M, Terano H, Kohsaka M, Aoki H, Imanaka H (1980) Studies on new phosphonic acid antibiotics. IV. Structure determination of FR-33289, FR-31564 and FR-32863. J Antibiot 33:29

224. Jomaa H, Wiesner J, Sanderbrand S, Altincicek B, Weidemeyer C, Hintz M, Turbachova I, Eberl M, Zeidler J, Lichtenthaler HK, Soldati D, Beck E (1999) Inhibitors of the nonmevalonate pathway of isoprenoid biosynthesis as antimalarial drugs. Science 285:1573
225. Brücher K, Illarionov B, Held J, Tschan S, Kunfermann A, Pein MK, Bacher A, Gräwert T, Maes L, Mordmüller B, Fischer M, Kurz T (2012) α-Substituted β-oxa isosteres of fosmidomycin: synthesis and biological evaluation. J Med Chem 55:6566
226. Fernandes JF, Lell B, Agnandji ST, Obiang RM, Bassat Q, Kremsner PG, Mordmüller B, Grobusch MP (2015) Fosmidomycin as an antimalarial drug: a meta-analysis of clinical trials. Future Microbiol 10:1375
227. Guinet J, Dvorak JA, Fujioka H, Keister DB, Muratova O, Kaslow DC, Aikawa M, Vaidya AB, Wellems TE (1996) A developmental defect in *Plasmodium falciparum* male gametogenesis. J Cell Biol 135:269
228. Campbell C, Collins W, Nguyen-Dinh P, Barber A, Broderson JR (1982) *Plasmodium falciparum* gametocytes from culture in vitro develop to sporozoites that are infectious to primates. Science 217:1048
229. Bhasin VK, Trager W (1984) Gametocyte-forming and non-gametocyte-forming clones of *Plasmodium falciparum*. Am J Trop Med Hyg 33:534
230. Walliker D, Quakyi I, Wellems T, McCutchan T, Szarfman A, London W, Corcoran L, Burkot T, Carter R (1987) Genetic analysis of the human malaria parasite *Plasmodium falciparum*. Science 236:1661
231. Ponnudurai T, Leeuwenberg AD, Meuwissen JH (1981) Chloroquine sensitivity of isolates of *Plasmodium falciparum* adapted to in vitro culture. Trop Geogr Med 33:50
232. Singh K, Agarwal A, Khan SI, Walker LA, Tekwani BL (2007) Growth, drug susceptibility, and gene expression profiling of *Plasmodium falciparum* cultured in medium supplemented with human serum or lipid-rich bovine serum albumin. J Biomol Screen 12:1109
233. Reed MB, Saliba KJ, Caruana SR, Kirk K, Cowman AF (2000) Pgh1 modulates sensitivity and resistance to multiple antimalarials in *Plasmodium falciparum*. Nature 403:906
234. Handunnetti SM, Hasler TH, Howard RJ (1992) *Plasmodium falciparum*-infected erythrocytes do not adhere well to C32 melanoma cells or CD36 unless rosettes with uninfected erythrocytes are first disrupted. Infect Immun 60:928
235. Jensen JB, Trager WP (1978) *Plasmodium falciparum* in culture: Establishment of additional strains. Am J Trop Med Hyg 27:743
236. Burkot TR, Williams JL, Schneider I (1984) Infectivity to mosquitoes of *Plasmodium falciparum* clones grown in vitro from the same isolate. Trans R Soc Trop Med Hyg 78:339
237. Thaithong S, Beale GH, Chutmongkonkul M (1983) Susceptibility of *Plasmodium falciparum* to five drugs: an in vitro study of isolates mainly from Thailand. Trans R Soc Trop Med Hyg 77:228
238. Sullivan JS, Sullivan JJ, Williams A, Grady KK, Bounngaseng A, Huber CS, Nace D, Williams T, Galland GG, Barnwell JW, Collins WE (2003) Adaptation of a strain of *Plasmodium falciparum* from Ghana to *Aotus lemurinus griseimembra, A. nancymai*, and *A. vociferans* monkeys. Am J Trop Med Hyg 69:593
239. Collins WE, Warren M, Skinner JC, Chin W, Richardson BB (1977) Studies on the Santa Lucia (El Salvador) strain of *Plasmodium falciparum* in *Aotus trivirgatus* monkeys. J Parasitol 63:52
240. Wootton JC, Feng X, Ferdig MT, Cooper RA, Mu J, Baruch DI, Magill AJ, Su X-Z (2002) Genetic diversity and chloroquine selective sweeps in *Plasmodium falciparum*. Nature 418:320
241. Siddiqui WA, Schnell JV (1973) Use of various buffers for in vitro cultivation of malarial parasites. J Parasitol 59:516
242. Chen P, Lamont G, Lamont TE, Kidson C, Brown G, Mitchell G, Stace J, Alpers M (1980) *Plasmodium falciparum* strains from Papua New Guinea: culture characteristics and drug sensitivity. SE Asian J Trop Med Public Health 11:435
243. Blunt JW, Copp BR, Keyzers RA, Munro MHG, Prinsep MR (2015) Marine natural products. Nat Prod Rep 32:116

244. Saraiva RG, Dimopoulos G (2020) Bacterial natural products in the fight against mosquito-transmitted tropical diseases. Nat Prod Rep 37:338
245. Tajuddeen N, van Heerden FR (2019) Antiplasmodial natural products: an update. Malar J 18:404
246. Schwikkard S, van Heerden FR (2002) Antimalarial activity of plant metabolites. Nat Prod Rep 19:675
247. Wright CW (2010) Recent developments in research on terrestrial plants used for the treatment of malaria. Nat Prod Rep 27:961
248. Wright CW (2015) Searching for new treatments of malaria. In: Heinrich M, Jäger AT (eds) Ethnopharmacology. Wiley-Blackwell, Chichester, UK, p 123
249. Pan WH, Xu XY, Shi N, Tsang SW, Zhang HJ (2018) Antimalarial activity of plant metabolites. Int J Mol Sci 19:1382
250. Bekono BD, Ntie-Kang F, Onguene PA, Lifongo LL, Sippl W, Fester K, Owono LCO (2020) The potential of anti-malarial compounds derived from African medicinal plants: a review of pharmacological evaluations from 2013 to 2019. Malar J 19:183
251. Silva GNS, Rezende LCD, Emery FS, Gosmann G, Gnoatto SCB (2015) Natural and semi-synthetic antimalarial compounds: emphasis on the terpene class. Mini Rev Med Chem 15:809
252. Kaur K, Jain M, Kaur T, Jain R (2009) Antimalarials from nature. Bioorg Med Chem 17:3229
253. Nair JJ, van Staden J (2019) Antiplasmodial lycorane alkaloid principles of the plant family Amaryllidaceae. Planta Med 85:637
254. Nair JJ, van Staden J (2019) The Amaryllidaceae as a source of antiplasmodial crinane alkaloid constituents. Fitoterapia 134:305
255. Morita H, Oshimi S, Hirasawa Y, Koyama K, Honda T, Ekasari W, Indrayanto G, Zaini NC (2007) Cassiarins A and B, novel antiplasmodial alkaloids from *Cassia siamea*. Org Lett 9:3691
256. Ekasari W, Widyawaruyanti A, Zaini NC, Syafruddin D, Honda T, Morita H (2009) Antimalarial activity of cassiarin A from the leaves of *Cassia siamea*. Heterocycles 78:1831
257. Rudyanto M, Tomizawa Y, Morita H, Honda T (2008) First total synthesis of cassiarin A, a naturally occurring potent antiplasmodial alkaloid. Org Lett 10:1921
258. Wang H, Cao F, Gao W, Wang X, Yang Y, Shi T, Wang Z (2021) Pd(II)-Catalyzed annulation reactions of epoxides with benzamides to synthesize isoquinolones. Org Lett 23:863
259. Gutierrez S, Coppola A, Sucunza D, Burgos C, Vaquero JJ (2016) Synthesis of 1-substituted isoquinolines by heterocyclization of TosM*IC* derivatives: total synthesis of cassiarin A. Org Lett 18:3378
260. Luesakul U, Palaga T, Krusong K, Ngamrojanavanich N, Vilaivan T, Puthong S, Muangsin N (2014) Synthesis, cytotoxicity, DNA binding and topoisomerase II inhibition of cassiarin A derivatives. Bioorg Med Chem Lett 24:2845
261. Morita H, Tomizawa Y, Deguchi J, Ishikawa T, Arai H, Zaima K, Hosoya T, Hirasawa Y, Matsumoto T, Kamata K, Ekasari W, Widyawaruyanti A, Wahyuni TS, Zaini NC, Honda T (2009) Synthesis and structure-activity relationships of cassiarin A as potential antimalarials with vasorelaxant activity. Bioorg Med Chem 17:8234
262. Presley CC, Du Y, Dalal S, Merino EF, Butler JH, Rakotonandrasana S, Rasamison VE, Cassera MB, Kingston DGI (2017) Isolation, structure elucidation, and synthesis of antiplasmodial quinolones from *Crinum firmifolium*. Bioorg Med Chem 25:4203
263. Kancharla P, Dodean RA, Li Y, Pou S, Pybus B, Melendez V, Read L, Bane CE, Vesely B, Kreishman-Deitrick M, Black C, Li Q, Sciotti RJ, Olmeda R, Luong TL, Gaona H, Potter B, Sousa J, Marcsisin S, Caridha D, Xie L, Vuong C, Zeng Q, Zhang J, Zhang P, Lin H, Butler K, Roncal N, Gaynor-Ohnstad L, Leed SE, Nolan C, Ceja FG, Rasmussen SA, Tumwebaze PK, Rosenthal PJ, Mu J, Bayles BR, Cooper RA, Reynolds KA, Smilkstein MJ, Riscoe MK, Kelly JX (2020) Lead optimization of second-generation acridones as broad-spectrum antimalarials. J Med Chem 63:6179
264. Du Y, Valenciano AL, Dai Y, Zheng Y, Zhang F, Zhang Y, Clement J, Goetz M, Kingston DGI, Cassera MB (2020) Anibamine and its analogues: potent antiplasmodial agents from *Aniba citrifolia*. J Nat Prod 83:569

265. François G, Timperman G, Eling W, Assi LA, Holenz J, Bringmann G (1997) Naphthylisoquinoline alkaloids against malaria: evaluation of the curative potentials of dioncophylline C and dioncopeltine A against *Plasmodium berghei* in vivo. Antimicrob Agents Chemother 41:2533
266. Schwedhelm KF, Horstmann M, Faber JH, Reichert Y, Bringmann G, Faber C (2007) The novel antimalarial compound dioncophylline C forms a complex with heme in solution. ChemMedChem 2:541
267. Bringmann G, Zhang G, Olschlager T, Stich A, Wu J, Chatterjee M, Brun R (2013) Highly selective antiplasmodial naphthylisoquinoline alkaloids from *Ancistrocladus tectorius*. Phytochemistry 91:220
268. Li J, Seupel R, Feineis D, Mudogo V, Kaiser M, Brun R, Brunnert D, Chatterjee M, Seo EJ, Efferth T, Bringmann G (2017) Dioncophyllines C_2, D_2, and F and related naphthylisoquinoline alkaloids from the Congolese liana *Ancistrocladus ileboensis* with potent activities against *Plasmodium falciparum* and against multiple myeloma and leukemia cell lines. J Nat Prod 80:443
269. Bringmann G, Messer K, Schwöbel B, Brun R, Assi LA (2003) Habropetaline A, an antimalarial naphthylisoquinoline alkaloid from *Triphyophyllum peltatum*. Phytochemistry 62:345
270. Xu M, Bruhn T, Hertlein B, Brun R, Stich A, Wu J, Bringmann G (2010) Shuangancistrotectorines A–E, dimeric naphthylisoquinoline alkaloids with three chiral biaryl axes from the Chinese plant *Ancistrocladus tectorius*. Chem Eur J 16:4206
271. Bringmann G, Lombe BK, Steinert C, Ioset KN, Brun R, Turini F, Heubi G, Mudogo V (2013) Mbandakamines A and B, unsymmetrically coupled dimeric naphthylisoquinoline alkaloids, from a Congolese *Ancistrocladus* species. Org Lett 15:2590
272. Tshitenge DT, Feineis D, Mudogo V, Kaiser M, Brun R, Seo EJ, Efferth T, Bringmann G (2018) Mbandakamine-type naphthylisoquinoline dimers and related alkaloids from the Central African liana *Ancistrocladus ealaensis* with antiparasitic and antileukemic activities. J Nat Prod 81:918
273. Fayez S, Li J, Feineis D, Ake Assi L, Kaiser M, Brun R, Anany MA, Wajant H, Bringmann G (2019) A near-complete series of four atropisomeric jozimine A_2-type naphthylisoquinoline dimers with antiplasmodial and cytotoxic activities and related alkaloids from *Ancistrocladus abbreviatus*. J Nat Prod 82:3033
274. Bringmann G, Zhang G, Büttner T, Bauckmann G, Kupfer T, Braunschweig H, Brun R, Mudogo V (2013) Jozimine A_2: the first dimeric Dioncophyllaceae-type naphthylisoquinoline alkaloid, with three chiral axes and high antiplasmodial activity. Chem Eur J 19:916
275. Lombe BK, Bruhn T, Feineis D, Mudogo V, Brun R, Bringmann G (2017) Cyclombandakamines A_1 and A_2, oxygen-bridged naphthylisoquinoline dimers from a Congolese *Ancistrocladus* liana. Org Lett 19:1342
276. Lombe BK, Feineis D, Mudogo V, Kaiser M, Bringmann G (2021) Spirombandakamine A_3 and cyclombandakamines A_8 and A_9, polycyclic naphthylisoquinoline dimers, with antiprotozoal activity, from a Congolese *Ancistrocladus* plant. J Nat Prod 84:1335
277. Moyo P, Shamburger W, van der Watt ME, Reader J, de Sousa ACC, Egan TJ, Maharaj VJ, Bringmann G, Birkholtz LM (2020) Naphthylisoquinoline alkaloids, validated as hit multistage antiplasmodial natural products. Int J Parasitol Drugs Drug Resist 13:51
278. Bouquet J, Rivaud M, Chevalley S, Deharo E, Jullian V, Valentin A (2012) Biological activities of nitidine, a potential anti-malarial lead compound. Malaria J 11:67
279. Gakunju DMN, Mberu EK, Dossaji SF, Gray AI, Waigh RD, Waterman PG, Watkins WM (1995) Potent antimalarial activity of the alkaloid nitidine, isolated from a Kenyan herbal remedy. Antimicrob Agents Chemother 39:2606
280. Kassim OO, Loyevsky M, Elliott B, Geall A, Amonoo H, Gordeuk VR (2005) Effects of root extracts of *Fagara zanthoxyloides* on the in vitro growth and stage distribution of *Plasmodium falciparum*. Antimicrob Agents Chemother 49:264
281. Rivaud M, Mendoza A, Sauvain M, Valentin A, Jullian V (2012) Short synthesis and antimalarial activity of fagaronine. Bioorg Med Chem 20:4856

282. Goodman CD, Austarheim I, Mollard V, Mikolo B, Malterud KE, McFadden GI, Wangensteen H (2016) Natural products from *Zanthoxylum heitzii* with potent activity against the malaria parasite. Malar J 15:481
283. Fadaeinasab M, Taha H, Fauzi PNM, Ali HM, Widyawaruyanti A (2015) Anti-malarial activity of isoquinoline alkaloids from the stem bark of *Actinodaphne macrophylla*. Nat Prod Commun 10:1541
284. Zahari A, Cheah F, Mohamad J, Sulaiman S, Litaudon M, Leong K, Awang K (2014) Antiplasmodial and antioxidant isoquinoline alkaloids from *Dehaasia longipedicellata*. Planta Med 80:599
285. Kanyanga CR, Munduku KC, Lumpu NS, Ehata TM, Bool-Miting MF, Kabangu KO, Maya MB, Cos P, Maes L, Vlietinck AJ, Tuenter E, Foubert K, Pieters L (2018) Isolation and structure elucidation of two antiprotozoal bisbenzylisoquinoline alkaloids from *Triclisia gilletii* stem bark. Phytochem Lett 28:19
286. Nonaka M, Murata Y, Takano R, Han Y, Kabir MHB, Kato K (2018) Screening of a library of traditional Chinese medicines to identify anti-malarial compounds and extracts. Malar J 17:244
287. Wright CW, Marshall SJ, Russell PF, Anderson MM, Phillipson JD, Kirby GC, Warhurst DC, Schiff PL (2000) In vitro antiplasmodial, antiamoebic, and cytotoxic activities of some monomeric isoquinoline alkaloids. J Nat Prod 63:1638
288. Teklemichael AA, Mizukami S, Toume K, Mosaddeque F, Kamel MG, Kaneko O, Komatsu K, Karbwang J, Huy NT, Hirayama K (2020) Anti-malarial activity of traditional Kampo medicine *Coptis* rhizome extract and its major active compounds. Malar J 19:204
289. Frederich M, Tits M, Angenot L (2008) Potential antimalarial activity of indole alkaloids. Trans R Soc Trop Med Hyg 102:11
290. Rosales PF, Bordin GS, Gower AE, Moura S (2020) Indole alkaloids: 2012 until now, highlighting the new chemical structures and biological activities. Fitoterapia 143:104558
291. Rocha e Silva LF, Montoia A, Amorim RCN, Melo MR, Henrique MC, Nunomura SM, Costa MRF, Andrade Netoe VF, Costa DS, Dantas G, Lavrado J, Moreira R, Paulo A, Pinto AC, Tadei WP, Zacardi RS, Eberlin MN, Pohlit AM (2012) Comparative in vitro and in vivo antimalarial activity of the indole alkaloids ellipticine, olivacine, cryptolepine and a synthetic cryptolepine analog. Phytomedicine 20:71
292. Onyeibor O, Croft SL, Dodson HI, Feiz-Haddad M, Kendrick H, Millington NJ, Parapini S, Phillips RM, Seville S, Shnyder SD, Taramelli D, Wright CW (2005) Synthesis of some cryptolepine analogues, assessment of their antimalarial and cytotoxic activities, and consideration of their antimalarial mode of action. J Med Chem 48:2701
293. Wright CW (2007) Recent developments in naturally derived antimalarials: cryptolepine analogues. J Pharm Pharmacol 59:899
294. Forkuo AD, Ansah C, Mensah KB, Annan K, Gyan B, Theron A, Mancama D, Wright CW (2017) In vitro anti-malarial interaction and gametocytocidal activity of cryptolepine. Malar J 16:496
295. Rickards RW, Rothschild JM, Willis AC, de Chazal NM, Kirk J, Kirk K, Saliba KJ, Smith GDS (1999) Calothrixins A and B, novel pentacyclic metabolites from *Calothrix* cyanobacteria with potent activity against malaria parasites and human cancer cells. Tetrahedron 55:13513
296. Becher PG, Beuchat J, Gademann K, Jüttner F (2005) Nostocarboline: isolation and synthesis of a new cholinesterase inhibitor from *Nostoc* 78–12A. J Nat Prod 68:1793
297. Barbaras D, Kaiser M, Brun R, Gademann K (2008) Potent and selective antiplasmodial activity of the cyanobacterial alkaloid nostocarboline and its dimers. Bioorg Med Chem Lett 18:4413
298. Rodrigues AL, Trachtmann N, Becker J, Lohanatha AF, Blotenberg J, Bolten CJ, Korneli C, de Souza Lima AO, Porto LM, Sprenger GA, Wittmann C (2013) Systems metabolic engineering of *Escherichia coli* for production of the antitumor drugs violacein and deoxyviolacein. Metab Eng 20:29
299. Wilkinson MD, Lai H-E, Freemont PS, Baum J (2020) A biosynthetic platform for antimalarial drug discovery. Antimicrob Agents Chemother 64:e02129

300. Fernandez LS, Buchanan MS, Carroll AR, Feng YJ, Quinn RJ, Avery VM (2009) Flinderoles A–C: antimalarial bis-indole alkaloids from *Flindersia* species. Org Lett 11:329
301. Robertson LP, Duffy S, Wang Y, Wang D, Avery VM, Carroll AR (2017) Pimentelamines A-C, indole alkaloids isolated from the leaves of the Australian tree *Flindersia pimenteliana.* J Nat Prod 80:3211
302. Sakai R, Higa T, Jefford CW, Bernardinelli G (1986) Manzamine A, a novel antitumor alkaloid from a sponge. J Am Chem Soc 108:6404
303. Zhou B-N, Slebodnick C, Johnson RK, Mattern MR, Kingston DGI (2000) New cytotoxic manzamine alkaloids from a Palaun sponge. Tetrahedron 56:5781
304. Rao KV, Kasanah N, Wahyuono S, Tekwani BL, Schinazi RF, Hamann MT (2004) Three new manzamine alkaloids from a common Indonesian sponge and their activity against infectious and tropical parasitic diseases. J Nat Prod 67:1314
305. Waters AL, Peraud O, Kasanah N, Sims JW, Kothalawala N, Anderson MA, Abbas SH, Rao KV, Jupally VR, Kelly M, Dass A, Hill RT, Hamann MT (2014) An analysis of the sponge *Acanthostrongylophora igens*' microbiome yields an actinomycete that produces the natural product manzamine A. Front Mar Sci 1:54
306. Peng J, Kudrimoti S, Prasanna S, Odde S, Doerksen RJ, Pennaka HK, Choo YM, Rao KV, Tekwani BL, Madgula V, Khan SI, Wang B, Mayer AM, Jacob MR, Tu LC, Gertsch J, Hamann MT (2010) Structure-activity relationship and mechanism of action studies of manzamine analogues for the control of neuroinflammation and cerebral infections. J Med Chem 53:61
307. Frédérich M, Jacquier M-J, Thepenier P, De Mol P, Tits M, Philippe G, Delaude C, Angenot L, Zeches-Hanrot M (2002) Antiplasmodial activity of alkaloids from various *Strychnos* species. J Nat Prod 65:1381
308. van Pelt-Koops JC, Pett HE, Graumans W, van der Vegte-Bolmer M, van Gemert GJ, Rottmann M, Yeung BKS, Diagana TT, Sauerwein RW (2012) The spiroindolone drug candidate NITD609 potently inhibits gametocytogenesis and blocks *Plasmodium falciparum* transmission to *Anopheles* mosquito vector. Antimicrob Agents Chemother 56:3544
309. Rottmann M, McNamara C, Yeung BKS, Lee MCS, Zou B, Russell B, Seitz P, Plouffe DM, Dharia NV, Tan J, Cohen SB, Spencer KR, Gonzalez-Paez GE, Lakshminarayana SB, Goh A, Suwanarusk R, Jegla T, Schmitt EK, Beck H-P, Brun R, Nosten F, Renia L, Dartois V, Keller TH, Fidock DA, Winzeler EA, Diagana TT (2010) Spiroindolones, a potent compound class for the treatment of malaria. Science 329:1175
310. Ashton TD, Devine SM, Mohrle JJ, Laleu B, Burrows JN, Charman SA, Creek DJ, Sleebs BE (2019) The development process for discovery and clinical advancement of modern antimalarials. J Med Chem 62:10526
311. Bouwman SA, Zoleko-Manego R, Renner KC, Schmitt EK, Mombo-Ngoma G, Grobusch MP (2020) The early preclinical and clinical development of cipargamin (KAE609), a novel antimalarial compound. Travel Med Infect Dis 36:101765
312. Spangenberg T, Burrows JN, Kowalczyk P, McDonald S, Wells TNC, Willis PA (2013) The open access malaria box: a drug discovery catalyst for neglected diseases. PLoS One 8:e62906
313. Bowman JD, Merino EF, Brooks CF, Striepen B, Carlier PR, Cassera MB (2014) Antiapicoplast and gametocytocidal screening to identify the mechanisms of action of compounds within the malaria box. Antimicrob Agents Chemother 58:811
314. Ghavami M, Merino EF, Yao ZK, Elahi R, Simpson ME, Fernandez-Murga ML, Butler JH, Casasanta MA, Krai PM, Totrov MM, Slade DJ, Carlier PR Cassera MB (2018) Biological studies and target engagement of the 2-*C*-methyl-D-erythritol 4-phosphate cytidylyltransferase (IspD)-targeting antimalarial agent (1*R*,3*S*)-MMV008138 and analogs. ACS Infect Dis 4:549
315. Gorki V, Walter NS, Singh R, Chauhan M, Dhingra N, Salunke DB, Kaur S (2020) β-Carboline derivatives tackling malaria: biological evaluation and docking analysis. ACS Omega 5:17993
316. Chierrito TPC, Aguiar ACC, de Andrade IM, Ceravolo IP, Gonçalves RAC, de Oliveira AJB, Krettli AU (2014) Anti-malarial activity of indole alkaloids isolated from *Aspidosperma olivaceum.* Malaria J 13:e142
317. Nugroho AE, Sugai M, Hirasawa Y, Hosoya T, Awang K, Hadi AH, Ekasari W, Widyawaruyanti A, Morita H (2011) New antiplasmodial indole alkaloids from *Hunteria zeylanica.* Bioorg Med Chem Lett 21:3417

318. Paciaroni NG, Perry DL II, Norwood IV, Murillo Solano C, Collins J, Tenneti S, Chakrabarti D, Huigens RW III (2020) Re-engineering of yohimbine's biological activity through ring distortion: identification and structure-activity relationships of a new class of antiplasmodial agents. ACS Infect Dis 6:159
319. Wright CW, Allen D, Cai Y, Chen Z, Phillipson JD, Kirby GC, Warhurst DC, Tits M, Angenot L (1994) Selective antiprotozoal activity of some *Strychnos* alkaloids. Phytother Res 8:149
320. Nugroho AE, Hirasawa Y, Piow WC, Kaneda T, Hadi AH, Shirota O, Ekasari W, Widyawaruyanti A, Morita H (2012) Antiplasmodial indole alkaloids from *Leuconotis griffithii*. J Nat Med 66:350
321. Frederich M, De Pauw M-C, Llabres G, Tits M, Hayette M-P, Brandt V, Penelle J, De Mol P, Angenot L (2000) New antimalarial and cytotoxic sungucine derivatives from *Strychnos icaja* roots. Planta Med 66:262
322. Beaufay C, Ledoux A, Jansen O, Bordignon A, Zhao S, Teijaro CN, Andrade RB, Quetin-Leclercq J, Frédérich M (2018) In vivo antimalarial and antitrypanosomal activity of strychnogucine B, a bisindole alkaloid from *Strychnos icaja*. Planta Med 84:881
323. Davis RA, Duffy S, Fletcher S, Avery VM, Quinn RJ (2013) Thiaplakortones A–D: antimalarial thiazine alkaloids from the Australian marine sponge *Plakortis lita*. J Org Chem 78:9608
324. Davis RA, Buchanan MS, Duffy S, Avery VM, Charman SA, Charman WN, White KL, Shackleford DM, Edstein MD, Andrews KT, Camp D, Quinn RJ (2012) Antimalarial activity of pyrroloiminoquinones from the Australian marine sponge *Zyzzya* sp. J Med Chem 55:5851
325. Lam CFC, Cadelis MM Copp BR (2020) Exploration of the electrophilic reactivity of the cytotoxic marine alkaloid discorhabdin C and subsequent discovery of a new dimeric C-1/N-13-linked discorhabdin natural product. Mar Drugs 18:404
326. Koeffli JB, Brockman JA, Moffat J (1950) The structure of febrifugine and isofebrifugine. J Am Chem Soc 72:3323
327. Coatney GR, Cooper WC, Culwell WB, White WC, Imboden Jr CA (1950) Studies in human malaria. XXV. Trial of febrifugine, an alkaloid obtained from *Dichroa febrifuga* Lour., against the Chesson strain of *Plasmodium vivax*. J Natl Malar Soc 9:183
328. Zhu S, Zhang Q, Gudise C, Wei L, Smith E, Zeng Y (2009) Synthesis and biological evaluation of febrifugine analogues as potential antimalarial agents. Bioorg Med Chem 17:4496
329. Zhu S, Chandrashekar G, Meng L, Robinson K, Chatterji D (2012) Febrifugine analogue compounds: synthesis and antimalarial evaluation. Bioorg Med Chem 20:927
330. Pines M, Spector I (2015) Halofuginone—the multifaceted molecule. Molecules 20:573
331. Fagbami L, Deik AA, Singh K, Santos SA, Herman JD, Bopp SE, Lukens AK, Clish CB, Wirth DF, Mazitschek R (2019) The adaptive proline response in *P. falciparum* is independent of PfeIK1 and eIF2alpha signaling. ACS Infect Dis 5:515
332. McLaughlin NP, Evans P, Pines M (2014) The chemistry and biology of febrifugine and halofuginone. Bioorg Med Chem 22:1993
333. Kikuchi H, Horoiwa S, Kasahara R, Hariguchi N, Matsumoto M, Oshima Y (2014) Synthesis of febrifugine derivatives and development of an effective and safe tetrahydroquinazoline-type antimalarial. Eur J Med Chem 76:10
334. Bhuyan BK (1962) Pactamycin production by *Streptomyces pactum*. Appl Microbiol 10:302
335. Otoguro K, Iwatsuki M, Ishiyama A, Namatame M, Nishihara-Tukashima A, Shibahara S, Kondo S, Yamada H, Omura S (2010) Promising lead compounds for novel antiprotozoals. J Antibiot 63:381
336. Feling RH, Buchanan GO, Mincer TJ, Kauffman CA, Jensen PR, Fenical W (2003) Salinosporamide A: a highly cytotoxic proteasome inhibitor from a novel microbial source, a marine bacterium of the new genus *Salinospora*. Angew Chem Int Ed 42:355
337. Prudhomme J, McDaniel E, Ponts N, Bertani S, Fenical W, Jensen P, Le Roch K (2008) Marine Actinomycetes: a new source of compounds against the human malaria parasite. PLoS One 3:e2335
338. Ito Y, Ohashi Y, Sakurai Y, Sakurazawa M, Yoshida H, Shigemi A, Okuda T (1968) New basic water-soluble antibiotics BD-12 and BY-81. II. Isolation, purification and properties. J Antibiot 21:307

339. Kawakami Y, Yamasaki K, Nakamura S (1981) The structures of component A_1 (=LL-AB664) and component A_2 (=LL-AC 541), streptothricin-like antibiotics. J Antibiot 34:921
340. Kubo M, Yatsuzuka W, Matsushima S, Harada K, Inoue Y, Miyamoto H, Matsumoto M, Fukuyama Y (2016) Antimalarial phenanthroindolizine alkaloids from *Ficus septica*. Chem Pharm Bull 64:957
341. Wasserman HH, Rodgers GC Keith DD (1969) Metacycloprodigiosin, a tripyrrole pigment from *Streptomyces longisporus ruber*. J Am Chem Soc 91:1263
342. Kim H-S, Hayashi M, Shibata Y, Wataya T, Mitamura T, Horii T, Kawauchi K, Hirata H, Tsuboi S, Moriyama Y (1999) Cycloprodigiosin hydrochloride obtained from *Pseudoalteromonas denitrificans* is a potent antimalarial agent. Biol Pharm Bull 22:532
343. Isaka M, Jaturapat A, Kramyu J, Tanticharoen M, Thebtaranonth Y (2002) Potent in vitro antimalarial activity of metacycloprodigiosin isolated from *Streptomyces spectabilis* BCC 4785. Antimicrob Agents Chemother 46:1112
344. Vitale GA, Sciarretta M, Palma Esposito F, January GG, Giaccio M, Bunk B, Spröer C, Bajerski F, Power D, Festa C, Monti MC, D'Auria MV, de Pascale D (2020) Genomics–metabolomics profiling disclosed marine *Vibrio spartinae* 3.6 as a producer of a new branched side chain prodigiosin. J Nat Prod 83:1495
345. Kancharla P, Li Y, Yeluguri M, Dodean RA, Reynolds KA, Kelly JX (2021) Total synthesis and antimalarial activity of 2-(*p*-hydroxybenzyl)-prodigiosins, isoheptylprodigiosin, and geometric isomers of tambjamine MYP1 isolated from marine bacteria. J Med Chem 64:8739
346. Ilias M, Ibrahim MA, Khan SI, Jacob MR, Tekwani BL, Walker LA, Samoylenko V (2012) Pentacyclic ingamine alkaloids, a new antiplasmodial pharmacophore from the marine sponge *Petrosid Ng5 Sp5*. Planta Med 78:1690
347. Xu M, Andrews KT, Birrell GW, Tran TL, Camp D, Davis RA, Quinn RJ (2011) Psammaplysin H, a new antimalarial bromotyrosine alkaloid from a marine sponge of the genus *Pseudoceratina*. Bioorg Med Chem Lett 21:846
348. Velten R, Erdelen C, Gehling M, Göhrt A, Gondol D, Lenz J, Lockhoff O, Wachendorff U, Wendisch D (1998) Cripowellin A and B, a novel type of Amaryllidaceae alkaloid from *Crinum powellii*. Tetrahedron Lett 39:1737
349. Presley CC, Krai P, Dalal S, Su Q, Cassera M, Goetz M, Kingston DGI (2016) New potently bioactive alkaloids from *Crinum erubescens*. Bioorg Med Chem 24:5418
350. Ferreira MC, Cantrell CL, Wedge DE, Goncalves VN, Jacob MR, Khan S, Rosa CA, Rosa LH (2017) Antimycobacterial and antimalarial activities of endophytic fungi associated with the ancient and narrowly endemic neotropical plant *Vellozia gigantea* from Brazil. Mem Inst Oswaldo Cruz 112:692
351. Cikos AM, Jurin M, Coz-Rakovac R, Jokic S, Jerkovic I (2019) Update on monoterpenes from red macroalgae: isolation, analysis, and bioactivity. Mar Drugs 17:537
352. Afolayan AF, Mann MG, Lategan CA, Smith PJ, Bolton JJ, Beukes DR (2009) Antiplasmodial halogenated monoterpenes from the marine red alga *Plocamium cornutum*. Phytochemistry 70:597
353. Schnermann MJ, Shenvi RA (2015) Syntheses and biological studies of marine terpenoids derived from inorganic cyanide. Nat Prod Rep 32:543
354. Angerhofer CK, Pezzuto JM, König GM, Wright AD, Sticher O (1992) Antimalarial activity of sesquiterpenes from the marine sponge *Acanthella klethra*. J Nat Prod 55:1787
355. White AM, Pierens GK, Skinner-Adams T, Andrews KT, Bernhardt PV, Krenske EH, Mollo E, Garson MJ (2015) Antimalarial isocyano and isothiocyanato sesquiterpenes with tri- and bicyclic skeletons from the nudibranch *Phyllidia ocellata*. J Nat Prod 78:1422
356. Wright AD, McCluskey A, Robertson MJ, MacGregor KA, Gordon CP, Guenther J (2011) Anti-malarial, anti-algal, anti-tubercular, anti-bacterial, anti-photosynthetic, and anti-fouling activity of diterpene and diterpene isonitriles from the tropical marine sponge *Cymbastela hooperi*. Org Biomol Chem 9:400
357. Pronin SV, Shenvi RA (2012) Synthesis of a potent antimalarial amphilectene. J Am Chem Soc 134:19604

358. Lu H-H, Pronin SV, Antonova-Koch Y, Meister S, Winzeler EA, Shenvi RA (2016) Synthesis of (+)-7,20-diisocyanoadociane and liver-stage antiplasmodial activity of the isocyanoterpene class. J Am Chem Soc 138:7268
359. Wright AD, Wang H, Gurrath M, König GM, Kocak G, Neumann G, Loria P, Foley M, Tilley L (2001) Inhibition of heme detoxification processes underlies the antimalarial activity of terpene isonitrile compounds from marine sponges. J Med Chem 44:873
360. Young RM, Adendorff MR, Wright AD, Davies-Coleman MT (2015) Antiplasmodial activity: the first proof of inhibition of heme crystallization by marine isonitriles. Eur J Med Chem 93:373
361. Aviles E, Prudhomme J, Le Roch KG, Rodriguez AD (2015) Structures, semisyntheses, and absolute configurations of the antiplasmodial alpha-substituted β-lactam monamphilectines B and C from the sponge *Svenzea flava*. Tetrahedron 71:487
362. Miyaoka H, Shimomura M, Kimura H, Yamada Y, Kim H-S, Yusuke W (1998) Antimalarial activity of kalihinol A and new relative diterpenoids from the Okinawan sponge, *Acanthella* sp. Tetrahedron 54:13467
363. Daub ME, Prudhomme J, Le Roch K, Vanderwal CD (2015) Synthesis and potent antimalarial activity of kalihinol B. J Am Chem Soc 137:4912
364. Daub ME, Prudhomme J, Ben Mamoun C, Le Roch KG, Vanderwal CD (2017) Antimalarial properties of simplified kalihinol analogues. ACS Med Chem Lett 8:355
365. Kyei-Baffour K, Davis DC, Boskovic Z, Kato N, Dai M (2020) Natural product-inspired aryl isonitriles as a new class of antimalarial compounds against drug-resistant parasites. Bioorg Med Chem 28:115678
366. Mohammad H, Kyei-Baffour K, Abutaleb NS, Dai M, Seleem MN (2019) An aryl isonitrile compound with an improved physicochemical profile that is effective in two mouse models of multidrug-resistant *Staphylococcus aureus* infection. J Glob Antimicrob Resist 19:1
367. Kyei-Baffour K, Mohammad H, Seleem MN, Dai M (2019) Second-generation aryl isonitrile compounds targeting multidrug-resistant *Staphylococcus aureus*. Bioorg Med Chem 27:1845
368. Nyaba ZN, Murambiwa P, Opoku AR, Mukaratirwa S, Shode FO, Simelane MBC (2018) Isolation, characterization, and biological evaluation of a potent anti-malarial drimane sesquiterpene from *Warburgia salutaris* stem bark. Malar J 17:296
369. Morita H, Mori R, Deguchi J, Oshimi S, Hirasawa Y, Ekasari W, Widyawaruyanti A, Hadi AH (2012) Antiplasmodial decarboxyportentol acetate and 3,4-dehydrotheaspirone from *Laumoniera bruceadelpha*. J Nat Med 66:571
370. Efange SM, Brun R, Wittlin S, Connolly JD, Hoye TR, McAkam T, Makolo FL, Mbah JA, Nelson DP, Nyongbela KD, Wirmum CK (2009) Okundoperoxide, a bicyclic cyclofarnesylsesquiterpene endoperoxide from *Scleria striatinux* with antiplasmodial activity. J Nat Prod 72:280
371. Shen Y, Cui SJ, Chen H, Shen L, Wang M, Dong X, Xiao CJ, Jiang B (2020) Antimalarial eudesmane sesquiterpenoids from *Dobinea delavayi*. J Nat Prod 83:927
372. Moon HI (2007) Antiplasmodial activity of ineupatorolides A from *Carpesium rosulatum*. Parasitol Res 100:1147
373. Chung IM, Kim MY, Moon HI (2008) Antiplasmodial activity of sesquiterpene lactone from *Carpesium rosulatum* in mice. Parasitol Res 103:341
374. Zhou B, Wu Y, Dalal S, Merino EF, Liu QF, Xu CH, Yuan T, Ding J, Kingston DGI, Cassera MB, Yue JM (2017) Nanomolar antimalarial agents against chloroquine-resistant *Plasmodium falciparum* from medicinal plants and their structure-activity relationships. J Nat Prod 80:96
375. Chukwujekwu JC, Smith P, Coombes PH, Mulholland DA, van Staden J (2005) Antiplasmodial diterpenoid from the leaves of *Hyptis suaveolens*. J Ethnopharmacol 102:295
376. Seephonkai P, Pyne SG, Willis AC, Lie W (2013) Bioactive compounds from the roots of *Strophioblachia fimbricalyx*. J Nat Prod 76:1358
377. Ebrahimi SN, Zimmermann S, Zaugg J, Smiesko M, Brun R Hamburger M (2013) Abietane diterpenoids from *Salvia sahendica*—antiprotozoal activity and determination of their absolute configurations. Planta Med 79:150

378. Gonzalez MA, Clark J, Connelly M, Rivas F (2014) Antimalarial activity of abietane ferruginol analogues possessing a phthalimide group. Bioorg Med Chem Lett 24:5234
379. Liu J, He X-F, Wang G-H, Merino EF, Yang S-P, Zhu R-X, Gan L-S, Zhang H, Cassera MB, Wang H-Y, Kingston DGI, Yue J-M (2014) Aphadilactones A–D, four diterpenoid dimers with DGAT inhibitory and antimalarial activities from a Meliaceae plant. J Org Chem 79:599
380. Yin JP, Gu M, Li Y, Nan FJ (2014) Total synthesis of aphadilactones A–D. J Org Chem 79:6294
381. Thiem DA, Sneden AT, Khan SI, Tekwani BL (2005) Bisnortriterpenes from *Salacia madagascariensis*. J Nat Prod 68:251
382. Du Y, Martin BA, Valenciano AL, Clement JA, Goetz M, Cassera MB, Kingston DGI (2020) Galtonosides A–E: antiproliferative and antiplasmodial cholestane glycosides from *Galtonia regalis*. J Nat Prod 83:1043
383. Muhammad I, Samoylenko V (2007) Antimalarial quassinoids: past, present and future. Expert Opin Drug Discov 2:1065
384. Li Z, Ruan J-Y, Sun F, Yan J-J, Wang J-L, Zhang Z-X, Zhang Y, Wang T (2019) Relationship between structural characteristics and plant sources along with pharmacology research of quassinoids. Chem Pharm Bull 67:654
385. Bertani S, Houel E, Stien D, Chevolot L, Jullian V, Garavito G, Bourdy G, Deharo E (2006) Simalikalactone D is responsible for the antimalarial properties of an Amazonian traditional remedy made with *Quassia amara* L. (Simaroubaceae). J Ethnopharmacol 108:155
386. Bertani S, Houel E, Jullian V, Bourdy G, Valentin A, Stien D, Deharo E (2012) New findings on simalikalactone D, an antimalarial compound from *Quassia amara* L. (Simaroubaceae). Exp Parasitol 130:341
387. Murakami N, Sugimoto M, Kawanishi M, Tamura S, Kim H-S, Begum K, Wataya Y, Kobayashi M (2003) New semisynthetic quassinoids with in vivo antimalarial activity. J Med Chem 46:638
388. Murakami N, Umezome T, Mahmud T, Sugimoto M, Kobayashi M, Wataya Y, Kim H-S (1998) Anti-malarial activities of acylated bruceolide derivatives. Bioorg Med Chem Lett 8:459
389. Lopatriello A, Sore H, Habluetzel A, Parapini S, D'Alessandro S, Taramelli D, Taglialatela-Scafati O (2019) Identification of a potent and selective gametocytocidal antimalarial agent from the stem barks of *Lophira lanceolata*. Bioorg Chem 93:103321
390. Chen M, Christensen SB, Blom J, Lemmich E, Nadelmann L, Fich K, Theander TG, Kharazmi A (1993) Licochalcone A, a novel antiparasitic agent with potent activity against human pathogenic protozoan species of *Leishmania*. Antimicrob Agents Chemother 37:2550
391. Chen M, Theander TG, Christenen SB, Hviid L, Zhai I, Kharazmi A (1994) Licochalcone A, a new antimalarial agent, inhibits in vitro growth of the human malaria parasite *Plasmodium falciparum* and protects mice from *P. yoelii* infection. Antimicrob Agents Chemother 38:1470
392. Iwatsuki M, Takada S, Mori M, Ishiyama A, Namatame M, Nishihara-Tsukashima A, Nonaka K, Masuma R, Otoguro K, Shiomi K, Omura S (2011) In vitro and in vivo antimalarial activity of puberulic acid and its new analogs, viticolins A–C, produced by *Penicillium* sp. FKI-4410. J Antibiot 64:183
393. Saito R, Sennari G, Nakajima A, Kimishima A, Iwatsuki M, Ishiyama A, Hokari R, Hirose T, Sunazuka T (2021) Discoveries and syntheses of highly potent antimalarial troponoids. Chem Pharm Bull 69:564
394. Harel D, Khalid SA, Kaiser M, Brun R, Wünsch B, Schmidt TJ (2011) Encecalol angelate, an unstable chromene from *Ageratum conyzoides* L.: Total synthesis and investigation of its antiprotozoal activity. J Ethnopharmacol 137:620
395. Harel D, Schepmann D, Prinz H, Brun R, Schmidt TJ, Wünsch B (2013) Enantioselective synthesis of encecaline-derived potent antimalarial agents. Org Biomol Chem 11:7342
396. Harel D, Schepmann D, Prinz H, Brun R, Schmidt TJ, Wünsch B (2013) Natural product derived antiprotozoal agents: Synthesis, biological evaluation and structure-activity relationships of novel chromene and chromane derivatives. J Med Chem 56:7442

397. Uth JF, Borgel F, Lehmkuhl K, Schepmann D, Kaiser M, Jabor VAP, Nonato MC, Krauth-Siegel RL, Schmidt TJ, Wunsch B (2021) Synthesis and biological evaluation of natural-product-inspired, aminoalkyl-substituted 1-benzopyrans as novel antiplasmodial agents. J Med Chem 64:6397
398. Kornsakulkarn J, Thongpanchang C, Chainoy R, Choowong W, Nithithanasilp S Thongpanchang T (2010) Bioactive metabolites from cultures of Basidiomycete *Favolaschia tonkinensis*. J Nat Prod 73:759
399. Kornsakulkarn J, Palasarn S, Choowong W, Thongpanchang T, Boonyuen N, Choeyklin R, Boonpratuang T, Isaka M, Thongpanchang C (2020) Antimalarial 9-methoxystrobilurins, oudemansins, and related polyketides from cultures of Basidiomycete *Favolaschia* species. J Nat Prod 83:905
400. Konziase B (2015) Protective activity of biflavanones from *Garcinia kola* against *Plasmodium* infection. J Ethnopharmacol 172:214
401. Oluwatosin A, Tolulope A, Ayokulehin K, Patricia O, Aderemi K, Catherine F Olusegun A (2014) Antimalarial potential of kolaviron, a biflavonoid from *Garcinia kola* seeds, against *Plasmodium berghei* infection in Swiss albino mice. Asian Pac J Trop Med 7:97
402. Gachet MS, Kunert O, Kaiser M, Brun R, Muñoz RA, Bauer R, Schühly W (2010) Jacaranone-derived glucosidic esters from *Jacaranda glabra* and their activity against *Plasmodium falciparum*. J Nat Prod 73:553
403. Xu L, He Z, Xue J, Chen X, Wei XJ (2010) β-Resorcylic acid lactones from a *Paecilomyces* fungus. J Nat Prod 73:885
404. Su Q, Dalal S, Goetz M, Cassera MB, Kingston DGI (2016) Antiplasmodial phloroglucinol derivatives from *Syncarpia glomulifera*. Bioorg Med Chem 24:2544
405. Harinantenaina L, Bowman JD, Brodie PJ, Slebodnick C, Callmander MW, Rakotobe E, Randrianaivo R, Rasamison VE, Gorka A, Roepe PD, Cassera MB, Kingston DG (2013) Antiproliferative and antiplasmodial dimeric phloroglucinols from *Mallotus oppositifolius* from the Madagascar dry forest. J Nat Prod 76:388
406. Eaton AL, Dalal S, Cassera MB, Zhao S, Kingston DG (2016) Synthesis and antimalarial activity of mallatojaponin C and related compounds. J Nat Prod 79:1679
407. Elbel KM, Guizzunti G, Theodoraki MA, Xu J, Batova A, Dakanali M Theodorakis EA (2013) A-ring oxygenation modulates the chemistry and bioactivity of caged *Garcinia* xanthones. Org Biomol Chem 11:3341
408. Ke H, Morrisey JM, Qu S, Chantarasriwong O, Mather MW, Theodorakis EA, Vaidya AB (2017) Caged *Garcinia* xanthones, a novel chemical scaffold with potent antimalarial activity. Antimicrob Agents Chemother 61:e01220-e1316
409. Dai Y, Harinantenaina L, Bowman JD, Da Fonseca IO, Brodie PJ, Goetz M, Cassera MB, Kingston DGI (2014) Isolation of antiplasmodial anthraquinones from *Kniphofia ensifolia*, and synthesis and structure-activity relationships of related compounds. Bioorg Med Chem 22:269
410. Bringmann G, Mutanyatta-Comar J, Maksimenka K, Wanjohi JM, Heydenreich M, Brun R, Muller WEG, Peter MG, Midiwo JO, Yenesew A (2008) Joziknipholones A and B: the frst dimeric phenylanthraquinones, from the roots of *Bulbine frutescens*. Chem Eur J 14:1420
411. Supong K, Thawai C, Suwanborirux K, Choowong W, Supothina S, Pittayakhajonwut P (2012) Antimalarial and antitubercular C-glycosylated benz[α]anthraquinones from the marine-derived *Streptomyces* sp. BCC45596. Phytochem Lett 5:651
412. Hayashi M, Yamada H, Mitamura T, Horii T, Yamamoto A, Moriyama Y (2000) Vacuolar H^+-ATPase localized in plasma membranes of malaria parasite cells, *Plasmodium falciparum*, is involved in regional acidification of parasitized erythrocytes. J Biol Chem 275:34353
413. Marchesini N, Vieira M, Luo S, Moreno SNJ, Docampo R (2005) A malaria parasite-encoded vacuolar H^+-ATPase is targeted to the host erythrocyte. J Biol Chem 280:36841
414. Shao CL, Linington RG, Balunas MJ, Centeno A, Boudreau P, Zhang C, Engene N, Spadafora C, Mutka TS, Kyle DE, Gerwick L, Wang, CY Gerwick WH (2015) Bastimolide A, a potent antimalarial polyhydroxy macrolide from the marine cyanobacterium *Okeania hirsuta*. J Org Chem 80:7849

415. Kinashi H, Someno K, Sakaguchi K (1984) Isolation and characterization of concanamycins A, B and C. J Antibiot 37:1333
416. Auparakkitanon S, Wilairat P (2006) Antimalarial activity of concanamycin A alone and in combination with pyronaridine. SE Asian J Trop Med Public Health 37:619
417. Hall TJ (1994) Cytotoxicity of vacuolar H^+-ATPase inhibitors to UMR-106 rat osteoblasts: an effect on iron uptake into cells? Cell Biol Internat 18:189
418. Ueda H, Nakajima H, Hori Y, Fujita T, Nishimura M, Goto T, Okuhara M (1994) FR901228, a novel antitumor bicyclic depsipeptide produced by *Chromobacterium violaceum* No. 968. I. Taxonomy, fermentation, isolation, physico-chemical and biological properties, and antitumor activity. J Antibiot 47:301
419. Engel JA, Jones AJ, Avery VM, Sumanadasa SDM, Ng SS, Fairlie DP, Adams TS, Andrews KT (2015) Profiling the anti-protozoal activity of anti-cancer HDAC inhibitors against *Plasmodium* and *Trypanosoma* parasites. Int J Parasitol Drugs Drug Resist 5:117
420. Saraiva RG, Huitt-Roehl CR, Tripathi A, Cheng YQ, Bosch J, Townsend CA Dimopoulos G (2018) *Chromobacterium* spp. mediate their anti-*Plasmodium* activity through secretion of the histone deacetylase inhibitor romidepsin. Sci Rep 8:6176
421. Annang F, Perez-Moreno G, Gonzalez-Menendez V, Lacret R, Perez-Victoria I, Martin J, Cantizani J, de Pedro N, Choquesillo-Lazarte D, Ruiz-Perez LM, Gonzalez-Pacanowska D, Genilloud O, Vicente F, Reyes F (2020) Strasseriolides A–D, a family of antiplasmodial macrolides isolated from the fungus *Strasseria geniculata* CF-247251. Org Lett 22:6709
422. Lane A, Stout EP, Lin AS, Prudhomme J, Le Roch K, Fairchild CR, Franzblau SG, Hay ME, Aalbersberg W, Kubanek J (2009) Antimalarial bromophycolides J–Q from the Fijian red alga *Callophycus serratus*. J Org Chem 74:2736
423. Stout EP, Cervantes S, Prudhomme J, France S, La Clair JJ, Le Roch K, Kubanek J (2011) Bromophycolide A targets heme crystallization in the human malaria parasite *Plasmodium falciparum*. ChemMedChem 6:1572
424. Nam SJ, Kauffman CA, Jensen PR, Fenical W (2011) Isolation and characterization of actinoramides A–C, highly modified peptides from a marine *Streptomyces* sp. Tetrahedron 67:6707
425. Williams DE, Dalisay DS, Patrick BO, Matainaho T, Andrusiak K, Deshpande R, Myers CL, Piotrowski JS, Boone C, Yoshida M, Andersen RJ (2011) Padanamides A and B, highly modified linear tetrapeptides produced in culture by a *Streptomyces* sp. isolated from a marine sediment. Org Lett 13:3936
426. Cheng KC, Cao S, Raveh A, MacArthur R, Dranchak P, Chlipala G, Okoneski MT, Guha R, Eastman RT, Yuan J, Schultz PJ, Su XZ, Tamayo-Castillo G, Matainaho T, Clardy J, Sherman DH, Inglese J (2015) Actinoramide A identified as a potent antimalarial from titration-based screening of marine natural product extracts. J Nat Prod 78:2411
427. Portmann C, Blom JF, Gademann K, Jüttner F (2008) Aerucyclamides A and B: isolation and synthesis of toxic ribosomal heterocyclic peptides from the cyanobacterium *Microcystis aeruginosa* PCC 7806. J Nat Prod 71:1193
428. Portmann C, Blom JF, Kaiser M, Brun R, Jüttner F, Gademann K (2008) Isolation of aerucyclamides C and D and structure revision of microcyclamide 7806A: heterocyclic ribosomal peptides from *Microcystis aeruginosa* PCC 7806 and their antiparasite evaluation. J Nat Prod 71:1891
429. Plouffe D, Brinker A, McNamara C, Henson K, Kato N, Kuhen K, Nagle A, Adrián F, Matzen JT, Anderson P, Nam T-G, Gray NS, Chatterjee A, Janes J, Yan SF, Trager R, Caldwell JS, Schultz PG, Zhou Y, Winzeler EA (2008) In silico activity profiling reveals the mechanism of action of antimalarials discovered in a high-throughput screen. Proc Natl Acad Sci USA 105:9059
430. Renner MK, Shen Y-C, Cheng X-C, Jensen PR, Frankmoelle W, Kauffman CA, Fenical W, Lobkovsky E, Clardy J (1999) Cyclomarins A–C, new antiinflammatory cyclic peptides produced by a marine bacterium (*Streptomyces* sp.). J Am Chem Soc 121:11273
431. Bürstner N, Roggo S, Ostermann N, Blank J, Delmas C, Freuler F, Gerhartz B, Hinniger A, Hoepfner D, Liechty B, Mihalic M, Murphy J, Pistorius D, Rottmann M, Thomas JR,

Schirle M, Schmitt EK (2015) Gift from nature: cyclomarin A kills mycobacteria and malaria parasites by distinct modes of action. ChemBioChem 16:2433
432. Keller-Schierlein W, Mihailovic ML, Prelog V (1959) Metabolic products of Actinomycetes. XV. The structure of echinomycin. Helv Chim Acta 42:305
433. Waring MJ, Wakelin MPG (1974) Echinomycin: a bifunctional intercalating antibiotic. Nature 252:653
434. Castillo U, Harper JK, Strobel GA, Sears J, Alesi K, Ford E, Lin J, Hunter M, Maranta M, Ge H, Yaver D, Jensen JB, Porter H, Robison R, Millar D, Hess WM, Condron M, Teplow D (2003) Kakadumycins, novel antibiotics from *Streptomyces* sp. NRRL 30566, an endophyte of *Grevillea pteridifolia*. FEMS Microbiol Lett 224:183
435. Taliaferro LG, Coulston F, Silverman M (1944) The antimalarial activity of tyrothricin against *Plasmodium gallinaceum*. J Infect Dis 75:179
436. Rautenbach M, Vlok NM, Stander M, Hoppe HC (2007) Inhibition of malaria parasite blood stages by tyrocidines, membrane-active cyclic peptide antibiotics from *Bacillus brevis*. Biochim Biophys Acta 1768:1488
437. van Epps HL (2006) René Dubos: unearthing antibiotics. J Exp Med 203:259
438. Gumila C, Ancelin J-M, Miquel G, Jeminet G, Delort A-M, Vial H (1996) Differential in vitro activities of ionophore compounds against *Plasmodium falciparum* and mammalian cells. Antimicrob Agents Chemother 40:602
439. David JM, Rajasekaran AK (2015) Gramicidin A: a new mission for an old antibiotic. J Kidney Cancer VHL 2:15
440. Gumila C, Ancelin ML, Delort AM, Jeminet G, Vial HJ (1997) Characterization of the potent in vitro and in vivo antimalarial activities of ionophore compounds. Antimicrob Agents Chemother 41:523
441. Yuan J, Cheng KC-C, Johnson RL, Huang R, Pattaradilokrat S, Liu A, Guha R, Fidock DA, Inglese J, Wellems TE, Austin CP, Su X-Z (2011) Chemical genomic profiling for antimalarial therapies, response signatures, and molecular targets. Science 333:724
442. Sweeney-Jones AM, Gagaring K, Antonova-Koch J, Zhou H, Mojib N, Soapi K, Skolnick J, McNamara CW, Kubanek J (2020) Antimalarial peptide and polyketide natural products from the Fijian marine cyanobacterium *Moorea producens*. Mar Drugs 18:167
443. Ciufolini MA, Lefranc D (2010) Micrococcin P1: structure, biology and synthesis. Nat Prod Rep 27:330
444. Rogers MJ, Cundliffe E, McCutchan TF (1998) The antibiotic micrococcin is a potent inhibitor of growth and protein synthesis in the malaria parasite. Antimicrob Agents Chemother 42:715
445. Degiacomi G, Personne Y, Mondesert G, Ge X, Mandava CS, Hartkoorn RC, Boldrin F, Goel P, Peisker K, Benjak A, Barrio MB, Ventura M, Brown AC, Leblanc V, Bauer A, Sanyal S, Cole ST, Lagrange S, Parish T, Manganelli R (2016) Micrococcin P1—a bactericidal thiopeptide active against *Mycobacterium tuberculosis*. Tuberculosis (Edinb) 100:95
446. D'Alessandro S, Corbett Y, Ilboudo DP, Misiano P, Dahiya N, Abay SM, Habluetzel A, Grande R, Gismondo MR, Dechering KJ, Koolen KM, Sauerwein RW, Taramelli D, Basilico N, Parapini S (2015) Salinomycin and other ionophores as a new class of antimalarial drugs with transmission-blocking activity. Antimicrob Agents Chemother 59:5135
447. Adovelande J, Schreve J (1996) Carboxylic ionophores in malaria chemotherapy: The effects of monensin and nigericin on *Plasmodium falciparum* in vitro and *Plasmodium vinckei petteri* in vivo. Life Sci 59:309
448. Scott PM, Van Walbeek W, MacLean WM (1971) Cladosporin, a new antifungal metabolite from *Cladosporium cladosporioides*. J Antibiot 24:747
449. Hoepfner D, McNamara CW, Lim CS, Studer C, Riedl R, Aust T, McCormack SL, Plouffe DM, Meister S, Schuierer S, Plikat U, Hartmann N, Staedtler F, Cotesta S, Schmitt EK, Petersen F, Supek F, Glynne RJ, Tallarico JA, Porter JA, Fishman MC, Bodenreider C, Diagana TT, Movva NR, Winzeler EA (2012) Selective and specific inhibition of the *Plasmodium falciparum* lysyl-tRNA synthetase by the fungal secondary metabolite cladosporin. Cell Host Microbe 11:654

450. Zhou J, Zheng L, Hei Z, Li W, Wang J, Yu B, Fang P (2020) Atomic resolution analyses of isocoumarin derivatives for inhibition of lysyl-tRNA synthetase. ACS Chem Biol 15:1016
451. Schulze CJ, Navarro G, Ebert D, DeRisi J, Linington RG (2015) Salinipostins A–K, long-chain bicyclic phosphotriesters as a potent and selective antimalarial chemotype. J Org Chem 80:1312
452. Yoo E, Schulze CJ, Stokes BH, Onguka O, Yeo T, Mok S, Gnadig NF, Zhou Y, Kurita K, Foe IT, Terrell SM, Boucher MJ, Cieplak P, Kumpornsin K, Lee MCS, Linington RG, Long JZ, Uhlemann A-C, Weerapana E, Fidock DA, Bogyo M (2020) The antimalarial natural product salinipostin A identifies essential α/β serine hydrolases involved in lipid metabolism in *P. falciparum* parasites. Cell Chem Biol 27:143
453. Zhang P, Nicholson DE, Bujnicki JM, Su X, Brendle JJ, Ferdig M, Kyle DE, Milhous WK, Chiang PK (2002) Angiogenesis inhibitors specific for methionine aminopeptidase 2 as drugs for malaria and leishmaniasis J Biomed Sci 9:34
454. Chen X, Xie S, Bhat S, Kumar N, Shapiro TA, Liu JO (2009) Fumagillin and fumarranol interact with *P. falciparum* methionine aminopeptidase 2 and inhibit malaria parasite growth in vitro and in vivo. Chem Biol 16:193
455. Mills B, Isaac RE, Foster R (2021) Metalloaminopeptidases of the protozoan parasite *Plasmodium falciparum* as targets for the discovery of novel antimalarial drugs. J Med Chem 64:1763
456. Fattorusso C, Campiani G, Catalanotti B, Persico M, Basilico N, Parapini S, Taramelli D, Campagnuolo C, Fattorusso E, Romano A, Taglialatela-Scafati O (2006) Endoperoxide derivatives from marine organisms: 1,2-dioxanes of the plakortin family as novel antimalaraial agents. J Med Chem 49:7088
457. Jimenez-Romero C, Ortiz I, Vicente J, Vera B, Rodrıguez AD, Nam S, Jove R (2010) Bioactive cycloperoxides isolated from the Puerto Rican sponge *Plakortis halichondrioides*. J Nat Prod 73:1694
458. Chianese G, Persico M, Yang F, Lin HW, Guo YW, Basilico N, Parapini S, Taramelli D, Taglialatela-Scafati O, Fattorusso C (2014) Endoperoxide polyketides from a Chinese *Plakortis simplex*: further evidence of the impact of stereochemistry on antimalarial activity of simple 1,2-dioxanes. Bioorg Med Chem 22:4572

David G. I. Kingston obtained a Ph.D. degree in chemistry under Lord Todd and Donald Cameron at Cambridge University in 1963. Subsequently, he held postdoctoral positions at the Massachusetts Institute of Technology and Cambridge University, before taking a faculty position at The State University of New York at Albany in 1966. In 1971, he moved to Virginia Tech, where he established an active research program on the chemistry of natural products, with a focus on the discovery and chemistry of anticancer and antimalarial natural products. One of the compounds he discovered is currently in Phase III clinical trials as the STAT3-targeted agent napabucasin. He was also one of the first chemists to investigate the chemistry of the anticancer drug taxol (paclitaxel). He is the author of over 400 journal articles as well as several book chapters, and is a co-editor of the book "Anticancer Agents from Natural Products". In addition to his scientific work, Prof. Kingston is a teaching elder at the Blacksburg Christian Fellowship. He retired from his position as a University Distinguished Professor at Virginia Tech in January 2020.

Maria Belen Cassera is an Associate Professor in the Department of Biochemistry and Molecular Biology and a faculty member of the Center for Tropical and Emerging Global Diseases (CTEGD) at the University of Georgia (UGA). She has a Doctor of Science degree in the area of the Biology of Host-Pathogen Interactions from the Department of Parasitology at the University of Sao Paulo in Brazil. She furthered her knowledge in the areas of enzymology and drug design as a postdoctoral trainee under the mentorship of Dr. Vern Schramm at the Albert Einstein College of Medicine in the Bronx, NY. Dr. Cassera started her independent career as an Assistant Professor in the Department of Biochemistry at Virginia Tech in January 2011, where she established strong metabolomics and drug discovery research programs. In August of 2016, she moved to UGA in Athens, Georgia, where her research focuses on two major areas: i) functional metabolomics and proteomics: for drug target identification and validation, and ii) drug discovery from synthetic and natural sources using both target-based and phenotypic screening. She has over 20 years of experience in the field of host-pathogen metabolic interactions of eukaryotic pathogens including *Plasmodium, Toxoplasma, Leishmania*, and *Trypanosoma*. She has published one book chapter in the topic of targeted metabolomics and over 70 research articles, where she has applied her knowledge in metabolomics and drug discovery.

Correction to: Antimalarial Natural Products

David G. I. Kingston and Maria Belen Cassera

Correction to:
Chapter "Antimalarial Natural Products" in: A. D. Kinghorn et al. (eds.), *Antimalarial Natural Products*, Progress in the Chemistryof Organic Natural Products 117, https://doi.org/10.1007/978-3-030-89873-1_1

The original version of the book was inadvertently published with an incorrect series of formulas (**197–199**) in the formula scheme on Page 70 of Chapter 1, which has now been replaced with the correct formula (**202**). The erratum chapter has been updated with the changes and the correct presentation is given here;

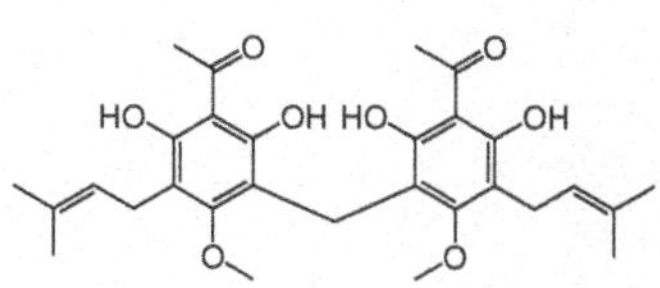

202 (mallatojaponin C)
Dd2 140 n*M*
A2780 1.3 μ*M*
SI 7.9

The updated version of this chapter can be found at
https://doi.org/10.1007/978-3-030-89873-1_1

A. D. Kinghorn, H. Falk, S. Gibbons, Y. Asakawa, J.-K. Liu, V. M. Dirsch (eds.),
Antimalarial Natural Products, Progress in the Chemistry
of Organic Natural Products 117, https://doi.org/10.1007/978-3-030-89873-1_2

The manufacturer's authorised representative in the EU is Springer Nature Customer Service Centre GmbH, Europaplatz 3, 69115 Heidelberg, Germany. If you have any concerns regarding our products, please contact ProductSafety@springernature.com

Printed and bound by CPI Group (UK) Ltd, Croydon, CR0 4YY
30/09/2026
02225262-0005